# A World Without Einstein

by

Robert L. Piccioni

Enjoy Exploring
Robert Piccioni

## A World Without Einstein

Almost all the products and technologies that touch our lives everyday depend on the discoveries of Albert Einstein, a man who experienced more failure and faced more rejection than most of us could overcome.

This book tells the story of Einstein's struggle to succeed, explains what he discovered, and shows how his achievements have enriched all our lives, both intellectually and financially. You will see why Albert Einstein truly deserved to be named "Person of the Century" by Time Magazine.

*A World Without Einstein* will be enjoyable and enlightening for both those who love science and those who fear physics or are allergic to math. Einstein's discoveries are explained in straight-forward English with helpful graphics and no complex math.

For more information about Robert Piccioni and his books, ebooks, DVDs, newsletters, and free online radio podcasts visit his website:

www.guidetothecosmos.com.

Real Science Publishing
3949 Freshwind Circle
Westlake Village, CA 91361 USA

Copyright © 2011 by Robert L. Piccioni
All rights reserved, including the right of reproduction in whole or in part, in any form.

Edited by Joan Piccioni
ISBN 978-0-9822780-4-8
Printed in USA
September, 2011

## About the Author

Dr. Robert L. Piccioni is a physicist, author, educator, and high-tech entrepreneur. He explains the wonders of science at high schools, universities, churches, civic organizations, business groups, astronomy clubs, adult education programs, and cruise ships.

Robert has a B.S. degree from Caltech where one of his professors was Nobel Laureate Richard Feynman, and a Ph.D. degree in high-energy physics from Stanford University where his thesis advisor was Nobel Laureate Melvin Schwartz. He was a member of the research faculty of Harvard University working with Nobel Laureate Carlo Rubbia.

Robert was principal of eight high-tech companies that produced CT-scanners, cancer therapy equipment, enabling technology for handicapped people, semiconductor fabrication machinery, bone densitometers, and endoscopic surgery devices. He holds several patents in medical, micro-electronic, and smart energy technologies.

Dr. Piccioni has lectured at Harvard, Stanford, Caltech, and UCLA. His presentations include: "Einstein for Everyone", "Stars, Energy, and Black Holes", "Our Universe: the Known and the Unknown", "Journey to the Dark Side: Dark Matter and Dark Energy", and "Can Life Be Merely An Accident?"

Robert is the author of two previous award-winning books:
***Everyone's Guide to Atoms, Einstein, and the Universe***
and ***Can Life Be Merely An Accident?***

# Table of Contents

# Dedicated to
# Luca Robert Piccioni

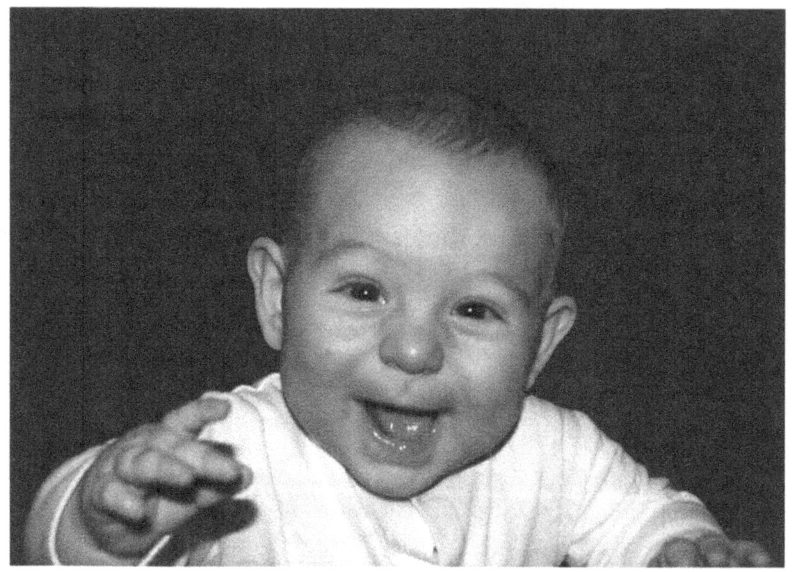

You are never too young or too old to wonder: Why?

I dedicate this book to my grandson Luca, whose natural curiosity inspires me. Children have an insatiable desire to explore and understand their world, which is the most important trait of a true scientist. Sometimes as we age, our natural curiosity is gradually submerged by the ocean of practical demands of "growing up." We should all occasionally emulate Luca and Einstein, who never outgrew his childish curiosity—it gave him joy throughout his life. I marvel to ponder what Luca will discover.

# Acknowledgments

I am delighted to thank Jason Piccioni, Luca's father, who helped develop this book's concept and make it more reader-friendly, and Joan Piccioni, Luca's grandmother, who originally proposed the book's subject and title, and was thereby obligated to do the editing, once again.

# 1

## Once Over Lightly

Almost everyone has heard of Albert Einstein—some say only Marilyn Monroe has greater worldwide name recognition. Most people know that Einstein was a great scientist, but many think his theories are so esoteric and purely academic that they have nothing to do with our daily lives.

Wrong!

In our technology-driven society, where even toothbrushes contain microcomputers, Einstein's legacy is everywhere. This book explores the incredible degree to which we all rely on his discoveries.

I designed this book to be enjoyable and enlightening for both those who love science and those who fear physics or are allergic to math. Each chapter describes in plain English something Einstein discovered and shows why this discovery is important to you. Also included are some sidebars—sections that explain the physics of Einstein's discoveries and are clearly titled "Physics..." A few meatier discussions are included in the appendices. You can simply skip the physics sections, if you wish—you won't miss anything that you'll need to enjoy the rest of the book. Although, after reading about the importance of Einstein's discoveries, I hope you will be intrigued and daring enough to go back and explore the fascinating science that makes all this possible. Those who wish to delve deeper into the science will enjoy my book ***Everyone's Guide to Atoms, Einstein, and the Universe.***

So, what do we owe to Einstein? For starters, how about:

TV
CDs
GPS
DVDs
Solar cells
Computers
Cell phones
Laser printers
Nuclear weapons
Bar code scanners
Digital electronics
Inertial navigation
Laser range finding
MRI body scanners
Automatic door openers
Laser eye and heart surgery
Magnetically levitated trains
Digital cameras and video cameras
Internet and telephone communication
Our most powerful telescopes and microscopes
And, our understanding of our place in the cosmos

That's quite a list—not too shabby for just one guy. Of course, Einstein didn't develop all these technologies all by himself. But all these technologies, and many more, *do* depend on his discoveries. As the above list clearly shows, Einstein's contributions not only advanced science but also enabled products and technologies that are vital to our economy—without Einstein we would all be poorer, both financially and intellectually.

What makes Einstein's accomplishments even more impressive is his personal story. Starting from his earliest days and continuing well into adulthood, Einstein failed at nearly everything he tried. No one had a clue that this seemingly hapless man would ever amount to much, let alone become the most famous scientist in history and Time Magazine's "Person of the Century."

Some claim that if Einstein hadn't made these discoveries, surely someone else would have…eventually. These same people might also say that if Christopher Columbus hadn't discovered the New World, eventually someone else would have. The second is probably true—as ship technology improved, crossing the Atlantic would have become ever less challenging, and someone would have done it, eventually. But how much longer might that have taken, how much less advanced might the United States be today, and how differently might world history have unfolded. Similarly, it may be true that someone else would eventually have made some of Einstein's lesser discoveries. But, how much longer would they have taken, how much slower would science and technology have advanced, and what would we be missing today as a result?

However, some of Einstein's discoveries are so extraordinary that without his genius they might remain undiscovered today, even after 100 years. Indeed, Einstein solved five critical problems that had stymied all the world's scientists for decades and even centuries. These were some of the most perplexing mysteries of science, and Einstein's solutions were profound and radical departures from long-cherished beliefs. Geniuses of his caliber are extremely rare—we haven't yet had another. We'll delve further into this question in the last chapter.

Let's end this first chapter with a list of Albert Einstein's major discoveries:

The existence of atoms
Light is both a wave and a particle
Special Relativity, replacing Newton's laws of motion
$E=mc^2$, equivalence and convertibility of mass and energy
General Relativity, replacing Newton's law of gravity
Theory for lasers

# 2

## "This Boy Will Never Amount To Anything"

Hermann and Pauline Einstein were married in 1876, and lived in the city of Ulm, in southwestern Germany. They were non-observant Jews who considered themselves more German than Jewish. In 1879, Pauline gave birth to their first child. They considered naming him Abraham, but decided that name sounded "too Jewish" and settled instead on Albert.

Albert had a rather inauspicious start in life. Even well into adulthood, there seemed little hope that he would ever succeed at anything, let alone become the most famous scientist in history. Hardly anyone has failed more often, or faced more rejection than did Albert Einstein.

As a child, Albert was very slow to learn to speak and was nicknamed "little dopey." Eventually people realized that young Albert was bright after all, and his poor communication skills weren't due to a lack of intelligence but perhaps to a lack of interest in interacting with others. Throughout his life, Einstein was internally focused—intensely absorbed with his own ponderings and his fascination with the beauty of nature. His focus was so intense that he would often "tune out" the world around him to concentrate on his own thoughts. This caused many people to assume he was absent-minded.

Above all else, Einstein prized personal freedom; he was contemptuous and defiant of authority. He proudly declared himself a "lone wolf", determined to live life on his own terms, with little concern for the opinions of others. He rebelliously declared: "Long live impudence" and "Blind respect for authority is the greatest enemy of truth."

A young Albert Einstein

Einstein despised the highly regimented German schools of the day, and his non-conformist spirit did not sit well with school authorities. Young Einstein was expelled from one school and asked to leave another. The headmaster sternly told Albert's father his little boy would never amount to anything. Amazingly, the headmaster was almost right.

When Einstein was 15 years old, his father's business suffered one of several reversals, and the family moved to Italy for better prospects. Albert was left behind with relatives in Germany to finish his last three years of high school. That didn't last long. The following autumn, Einstein dropped out of high school. Some say the school administration pushed him out. In any case, the authorities had no regrets at Einstein's departure. He rejoined his family in Italy, and renounced his German citizenship.

Einstein sought admission to the Polytechnic Institute in Zurich, Switzerland, which was then considered a science and math teacher's college. It was not in the top tier of European universities, and had no Ph.D. programs. However, Einstein failed the entrance exam, with particularly poor marks in French.

Advised to finish high school and reapply the following year, Einstein enrolled in a school in the nearby Swiss village of Aarau. This school's approach, based on the educational philosophy of Johann Pestalozzi, was a radical departure from what Einstein had experienced in Germany. To his delight, they emphasized independent thinking, visualization, and nurturing each student's "inner child." Einstein hardly needed encouragement in independent thinking, but the emphasis on visualization may have had a very positive and lasting benefit. Einstein said that he rarely thought in words, "A thought comes, and I may try to express it in words afterwards."

It was in Aarau that Einstein had the first thoughts that ultimately led to his Theory of Special Relativity. He tried to imagine what he would see if he could run alongside a light beam at the speed of light. He reasoned that the light beam should appear stationary, but how could a wave exist without motion? It took Einstein ten years to resolve this dilemma.

The next year, he reapplied to the Polytechnic and was admitted. Initially, all went well for Einstein. At the end of his second year, he ranked first in his class, but gradually, Einstein's insolence and disrespect of authority became issues again. Einstein habitually called his physics professor "Herr Weber" instead of the proper honorific "Herr *Professor* Weber"—a deliberate and seemingly pointless affront. Einstein skipped classes he found boring and ignored assignments he considered beneath his great intellect. His mathematics professor called him a "lazy dog." He even failed a physics course, *Physical Experiments for Beginners*.

In 1900, Einstein graduated from the Polytechnic, fifth in a class of six—next to last. With low marks and horrendous references, particularly from "Herr Weber", doors closed in his face all across Europe. For two years, Einstein applied for jobs at every university and technical institute, from Norway to Italy, and was summarily rejected—he was never even offered an interview. He was the only member of his graduating class who wasn't offered professional employment. Had he not received financial support from relatives, Einstein might have ended up living on the streets.

Largely through his own actions, Einstein had placed himself firmly on a road to failure and obscurity. It seemed he might well fulfill the predictions of his many detractors and never amount to anything.

Finally, in 1902, the influential father of Einstein's close friend, Marcel Grossman, pulled some strings in the Swiss government. Funds were allocated for a new entry-level position in the patent office in Bern. The job description was tailored to match Einstein's meager credentials. He applied, but failed the interview—Albert Einstein was rated technically unqualified to be a government clerk. Fortunately for science, the quest for competent civil servants was trumped by political influence, and Einstein was hired as patent clerk 3rd class, a minimum wage job at the bottom of the totem pole.

Einstein submitted a Ph.D. thesis to the University of Zurich, hoping a doctorate degree would improve his career prospects. But in his thesis he sharply criticized leading scientists of the day and his thesis was rejected.

In late 1902, at the nadir of Einstein's life, his father died. Hermann never lived to see his son achieve even a glimmer of the success and celebrity that later came to the world's most famous scientist. Albert was profoundly despondent at his father's death, perhaps because they had never forged the bond that he so desired.

It is interesting to note the scientific climate in which Einstein hoped to succeed. At the end of the 19th century, many felt that science and technology had reached the limits of what was possible. One of the world's leading physicists, Lord Kelvin, gave the keynote speech at the centennial celebration of Britain's Royal Society, the elite of British science, and said: "There is nothing new to be discovered in physics."

In 1899, the head of the U.S. Patent Office is said to have recommended closing the patent office because he believed: "Everything that can be invented, has been invented."

At about the same time, Henri Poincare, the leading French mathematician and physicist, said there were only three interesting problems left in physics: Brownian motion, the photoelectric effect, and the search for luminiferous ether. These were seen more as nagging loose ends than as opportunities for great advances.

Isn't it amazing that the leading scientists of the day had no clue that the greatest scientific discoveries in history were just around the corner?

Desperate to resurrect his failed career and facing a dearth of research topics, Einstein determined he must solve all three of Poincare's problems, and do so in spectacular style.

Einstein did indeed succeed, due to sheer genius and tremendous persistence. Despite an endless series of failures and rejections, Einstein never gave up on himself or his dreams. Asked later in life why he had discovered so much more than others, he answered: "It's not that I'm so smart, it's that I stay with problems longer." Einstein pondered some problems nearly non-stop for ten years or more, even when there was no glimmer of hope, and yet he never quit trying. Such dedication is truly remarkable, particularly in the present day, when for many it seems instant gratification doesn't come soon enough.

In 1905, several ideas that Einstein had been working on for as long as ten years gelled—he published five spectacular papers that solved all of Poincare's problems and completely revolutionized physics and mankind's understanding of the world around us. Thus, 1905 is called Einstein's ***Miracle Year***, which is a bit of a misnomer, because he actually published all five papers in only seven months. Any one of these papers could have earned a Nobel Prize. No other scientist has ever come close to matching this achievement.

Understandably, Einstein believed his spectacular papers would soon result in spectacular personal success, with universities competing for him with glorious job offers. But, again he was rejected, or at least ignored. He received no congratulations, no invitations to present his work, and no offers of employment. After two years of futility and frustration, Einstein applied to become a high school physics teacher, and was rejected.

It was only in 1909 that Einstein received the first small measure of professional success, a junior professorship at the University of Zurich. He was then 30 years old and had been a patent clerk for 7 years. It's not far off to say that for his first 30 years, Einstein failed at nearly everything he tried and was rejected by nearly everyone. Many would have given up long before—only because Einstein never quit do we all know his name today.

# 3

# Got Atoms?

For 2500 years, people wondered: "What is everything really made of?" Now we think we know.

The material in everything we see around us seems smooth and continuous. When we pour a glass of water, we don't see chunks of water tumble out; we see a continuous stream. For centuries, many people believed matter remained continuous, even down to the very smallest dimensions. But others believed matter was ultimately made of discrete parts that were too small for us to see.

Imagine examining a long, thin wire under a microscope, as illustrated in Figure 3.1. If matter were continuous, the wire would look smooth and featureless, even at the most extreme magnification. We could cut the wire into smaller and smaller pieces, and even the smallest piece would still retain the essence of being wire. But, if matter is ultimately made of discrete parts, at sufficient magnification, a wire might look like a chain made of individual links. Chains can have only certain lengths; they can be 2 links long, or 200 links long, but never 2½ links long. And unlike a wire, if a chain is cut into very small pieces, the pieces are no longer chain-like—the essential character of being a chain is lost once its links are cut.

The Greek philosopher Democritus is credited with being the first to name the small discrete pieces from which he believed all material

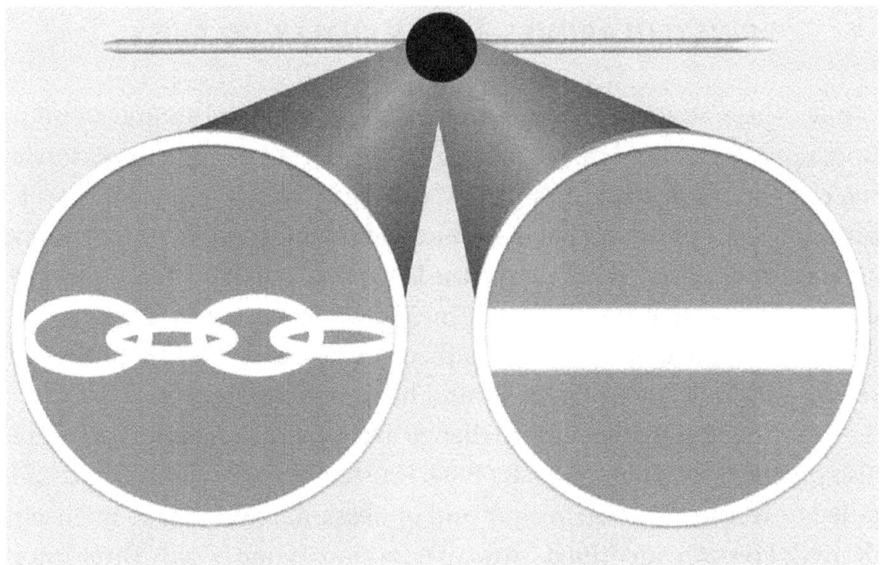

Figure 3.1. How would a wire look in a microscope, at extreme magnification? Would it look smooth and featureless, as on the right, or would we see small discrete pieces that are linked together, as on the left?

objects are made. He called them ***atomos***, a Greek word meaning uncuttable. The "atomic" debate continued for millennia as an entirely philosophical discussion. Without an appropriate microscope, no one could conclusively prove or disprove the existence of atoms.

In the $19^{th}$ century, physicists developed Thermodynamics, a theory of heat. It proved very effective in describing the behavior of gases and in designing efficient steam engines. Thermodynamics deals with bulk properties of matter, such as volume, pressure, and temperature. Ludwig Boltzmann, an Austrian mathematical physicist, sought a deeper understanding of why Thermodynamics worked so well. Using Newton's laws of mechanics and the mathematics of probability and statistics, Boltzmann developed a successful explanation of Thermodynamics based on the assumption that gases are composed of an immense number of very small atoms.

Despite the success of Boltzmann's theory, many scientists challenged him and vehemently attacked his atomic assumption as unproven and unjustified. They demanded direct evidence: "If atoms exist, show me one."

## EINSTEIN PROVES ATOMS REALLY DO EXIST

While others attacked Boltzmann for his "unjustified" atomic assumption, Einstein focused on something he considered of great importance, the elegance of Boltzmann's theory, calling it "absolutely magnificent." Einstein had a profound belief in the fundamental simplicity, harmony, and beauty of nature. He believed the laws of nature that science seeks to discover are simple and elegant. Einstein was convinced that the mathematical elegance of Boltzmann's theory was the result of a profound reality—the existence of atoms—and he wanted to prove it.

Certainly the intellectual challenge excited Einstein, but there were also practical considerations. By 1904, Einstein was 25 years old and had failed to secure the academic job and professional status he so intensely desired. Poincare identified Brownian motion as one of only three interesting problems remaining to be solved in physics. To resurrect his failed career, Einstein needed a dramatic success, which he achieved by solving the 78-year-old mystery of Brownian motion.

In 1827, English botanist Robert Brown used a microscope to study minute pollen grains suspended in liquid. For no apparent reason, the pollen grains continuously and randomly jittered this way and that. They moved one way and then would suddenly change direction, as shown in Figure 3.2. Brown wasn't looking for exciting new physics. He wanted to study pollen and was annoyed that the grains kept jittering. To him, "Brownian" motion was a nuisance.

Some scientists suggested these jitters might be due to collisions between the pollen grains and atoms that were too small to be seen. But their suggestions weren't quantitative and no one had a comprehensive theory to explain the observations. Additionally, the observations themselves were of little value since the motion of the pollen grains was so erratic. Thus, it was impossible to reach any definite conclusions about what caused Brownian motion, and the atomic debate continued.

Einstein conceived a more effective way to analyze these motions. Sometimes it is better to focus on the forest rather than on individual trees. Einstein realized that measuring the result of many collisions would be easier and more precise than trying to deal with each jitter separately.

Figure 3.2. When viewed under a microscope, a pollen grain is continually moving, jittering this way and that for no apparent reason. It might start at the left arrow and traverse the erratic path shown above, ending a minute later at the right arrow.

In 1905, he published a ***diffusion*** equation that predicts the average distance pollen grains travel in a given time. That distance depends on properties of the pollen and the liquid, which are easily measured, and also on the number of atoms in a given amount of liquid.

Meticulous experiments by French physicist Jean Baptiste Perrin confirmed the predictions of Einstein's equation, and enabled the first definitive measurements of the mass and size of atoms. Einstein had found the solution. The 78-year mystery of Brownian motion and the 2500-year atomic debate were finally settled.

Everything we see really is made of atoms.

Figure 3.3. With our most advanced technologies, we can now image individual atoms. Each dot above is a single xenon atom, imaged by an atomic force microscope in the laboratory of a large, well-known, international corporation.

## WHAT MAKES ATOMS TICK?

Atoms are very small by human standards. One billion carbon atoms lined up in a row would be about one foot long. A 130-pound person contains 6 billion, billion, billion atoms. As small as atoms are, we now know they are not the smallest components of matter. Atoms can indeed be cut into smaller pieces, and some of those pieces can be cut even further.

If we looked inside an atom we would see the image in Figure 3.4. At first, atoms seem to have only two parts: an inside and an outside. The inside is the ***nucleus***, which has a positive electric charge. The outside is a "cloud" of ***electrons***, which has a negative electric charge. Opposite charges (positive and negative) attract one another, thus holding atoms together.

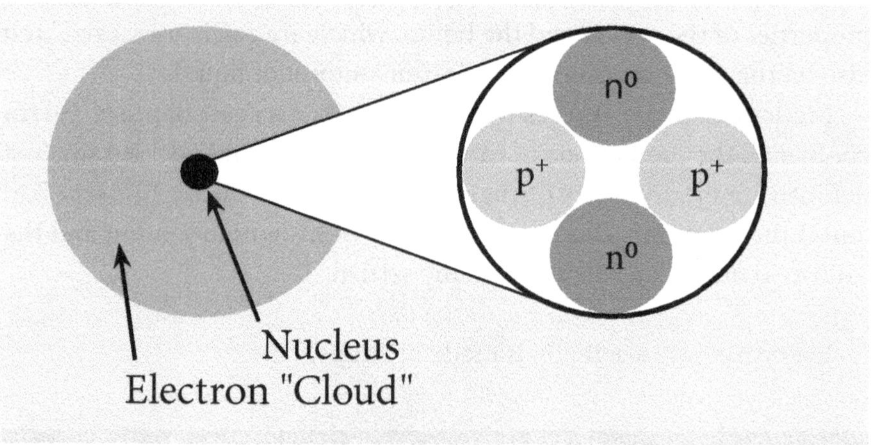


Figure 3.4. Atoms have two parts: a nucleus surrounded by a much larger cloud of electrons. Probing even deeper, the nucleus (expanded on the right) contains protons ($p^+$) and neutrons ($n^0$).

The electron cloud is 100,000 times larger than the nucleus. If one enlarged an atom until it filled a football stadium, the nucleus would be the size of the head of a pin. But even though it is so much smaller, the nucleus dominates the atom. Compared to the electrons, the nucleus typically has 4000 times more mass, and can contain a million times more extractable energy.

Electrons don't seem to have any internal parts, but nuclei do—they are composed of ***protons***, which have a positive electric charge, and ***neutrons***, which have zero charge. Each nucleus must have at least one proton to provide a positive charge to attract electrons. The number of protons in an atom's nucleus is its most important characteristic: its ***element number***. There are over 100 elements in the ***Periodic Table***, each with different properties. Gold is element number 79 with 79 protons in its nucleus. Lead is element number 82 with 3 more protons than gold. What a difference 3 little protons make.

The number of neutrons is generally similar to the number of protons, but it does vary. Two atoms with the same number of protons and different numbers of neutrons are the same element and have the same chemical properties, but are different ***isotopes***.

The most common atoms have relatively small nuclei and many of these have equal numbers of protons and neutrons. Carbon has 6 of each, nitrogen 7 of each, and oxygen 8 of each. In larger nuclei, neutrons outnumber protons. Iron has 30 neutrons and 26 protons, while uranium has 146 neutrons and 92 protons, in their most common isotopes.

Atoms generally have the same number of protons and electrons, making them neutral—having zero net electric charge. Atoms with different numbers of protons and electrons have a non-zero electric charge; these are called ***ions***.

## DIVERSE THINGS THAT ATOMS MAKE

Normal hydrogen has the smallest nucleus: one proton and no neutrons. While it is the smallest atom, hydrogen accounts for 92% of all atoms and 74% of all the atomic mass in the universe. Helium is element number 2; its normal isotope has two protons and two neutrons. Helium accounts for 7.5% of the atoms and 24% of the mass. Together, hydrogen and helium make up almost everything we see in the cosmos, accounting for 99.8%

of all atoms and 98% of all atomic mass. Of the remaining mass, 1% is oxygen, 0.5% is carbon, and all other elements are much less abundant.

On Earth, atomic abundances are quite different. Earth has almost no helium and very little hydrogen; it is primarily composed of iron (32%), oxygen (30%), silicon (15%), and magnesium (14%).

The atomic composition of the human body is very different from that of the universe, the stars, and the Earth. By mass, we are 65% oxygen, 18% carbon, 10% hydrogen, 3% nitrogen, 1.5% calcium, and 2.5% other elements. Life is made of fine and rare ingredients.

| Abundance of Elements by Mass | | | |
|---|---|---|---|
| Element | Human | Earth | Universe |
| Hydrogen, H | 10% | <0.01% | 74% |
| Helium, He | 0 | 0 | 24% |
| Oxygen, O | 65% | 30% | 1% |
| Carbon, C | 18% | <0.05% | 0.5% |
| Iron, Fe | <0.05% | 32% | 0.1% |

## WHAT WE GAIN BY UNDERSTANDING ATOMS

Richard Feynman, recipient of the 1965 Nobel Prize in Physics, was one of the most important and influential scientists of the 20$^{th}$ century. He was also one of my professors at Caltech, and a friend of my father, physicist Oreste Piccioni. Feynman taught me Quantum Electrodynamics and how to shoot pool. He once said that the greatest achievement of science, ever, was the discovery of atoms. He went on to say that if mankind ever contacts intelligent life on a distant world and we can send them only a single sentence to demonstrate our scientific prowess, that sentence should be: "Everything is made of atoms."

Almost every field of science is based on understanding atoms, their components, and their interactions. For example, the ways electrons orbit the nuclei of various atoms determine which chemical reactions are possible and how these reactions will proceed. With powerful computers, chemists can calculate properties of various compounds and thereby accelerate development of new products.

Molecular biologists use their knowledge of atoms, and how atoms combine in molecules, to guide their research. They are discovering why certain molecules can penetrate cell walls and others can't, how the vital processes of life function, and what impact various medicines, toxins, and pathogens have on different types of cells.

Understanding the properties of carbon and oxygen isotopes has even allowed scientists to determine the body temperature of dinosaurs that died 100 million years ago.

Similarly, understanding the electronic properties of atoms enabled the development of microelectronic devices, which comprise the guts of computers and the enormous variety of digital tools that are drastically changing our lives, as we'll discuss further in chapter 11.

The properties of microscopic atoms even help us comprehend the vast universe that surrounds us. By understanding atoms, we have discovered the age of the Earth, the age and size of our universe, what it's made of, how it began, and how it will end.

All this is contained within one of nature's smallest objects. That tiny atoms open so many windows demonstrates how intricately interconnected nature is.

# 4

# Einstein Sees The Light

In 1905, Einstein solved the mystery of the photoelectric effect—how light can make electricity, sometimes. Using Einstein's discovery, scientists and engineers have developed:

digital cameras
video cameras
TV cameras
solar cells
night vision goggles
automatic door openers
safety devices for garage doors
devices for endoscopic surgery
and the detectors for our most advanced telescopes,
including the Hubble Space Telescope.

## PHYSICS OF PHOTOELECTRIC EFFECT

As early as 1839, physicists observed that light hitting a metal surface causes an electric current, ***sometimes***. Over the next 66 years, many careful experiments led to the conclusion that the electric current was caused by light knocking electrons out of the metal. But a great mystery remained—why only ***sometimes***.

Before Einstein, everyone was absolutely sure there were two separate entities in nature: particles and waves. And everyone was also sure that light was a wave, because it did all the things that good waves do, including diffraction and interference. Knocking an electron out of a metal takes energy. Electrons have negative electric charge and are attracted to the metal's atomic nuclei, which have positive electric charge. Thus, it takes energy to overcome that electric attraction and push an electron away from the metal's nuclei. That's not a problem because waves can pack lots of energy—just ask surfers. Hence, a bright enough light should easily push over a few puny electrons and eject them from the metal.

All this made sense, but it turned out that nature doesn't work that way. Experiments found that red light, no matter how bright, didn't eject electrons from typical metals. However, blue light always ejected electrons. Even when the blue light delivered much less energy than the red light, blue succeeded where red failed. What was wrong with red?

In 1905, Einstein solved this mystery by proclaiming the unthinkable—light is ***both*** a wave and a particle—two things that everyone else was absolutely sure were complete opposites. Einstein said that a beam of light is actually a stream of an enormous number of individual particles called ***photons***.

Knowing that, let's consider the fate of a single electron. Even with the brightest light, Einstein said, an electron can be hit by only one photon at a time—no double-teaming. The electron will be ejected only if that single photon has enough energy to overcome its electric attraction to the metal's nuclei. The photon's energy is directly proportional to its frequency. Blue light has higher frequency than red light, and thus a photon of blue light has higher energy than a red light photon. Blue photons are energetic enough to eject electrons from typical metals and red ones aren't. A lot of little nudges by red photons just won't do; it takes one big whack from a single blue photon to knock an electron loose.

## LIGHT DETECTORS EXTEND OUR VISION

Without understanding the underlying science, it is very difficult, often impossible, to make a new technology useful. To optimize performance one needs to know all the pieces of the puzzle and how they fit together. Understanding the photoelectric effect, thanks to Einstein, enabled the development of countless devices that enrich society.

Photoelectric light detectors are used in a large variety of safety devices. If your car sticks out a bit, your garage door doesn't smash into it because a light beam shines across the doorway and is detected by a photoelectric sensor. If the sensor "sees" the light, it's OK to close the door. But if your car, your cat, or God forbid your young child is in the doorway blocking the light beam, the sensor prevents the door from closing. Similar systems protect factory workers from dangerous machinery by placing light beams between them and moving parts or other hazards. These devices are very simple and cheap to implement, and save lives and valuable property.

The photoelectric effect is also the core technology of all electronic imaging systems—everything that doesn't use film. A digital still camera focuses light onto a CCD (Charge Coupled Device) detector, which is an array of microelectronic devices that convert light into electric charge using the photoelectric effect. After the shutter closes, the charge in each device of the array is measured and recorded on the memory stick. Digital video cameras do the same thing but at much higher processing speeds.

Night vision systems also utilize the photoelectric effect with special materials that work at much lower light frequencies than metals. These systems are much more sensitive than our eyes and can detect infrared light that our eyes cannot. The frequency and energy of infrared light is lower than red light, which is the lowest frequency light our eyes can see. All warm objects, including people, emit infrared. So, even in "total darkness", night vision systems can image the ***heat signatures*** of people and vehicles, which is extremely useful for soldiers, police, and rescue workers.

Do you watch TV? If so, thank Einstein, because TV cameras, both analog and digital, use the photoelectric effect to record the images we see of news, sports, and our favorite comedies and dramas. No Einstein, no Tony Soprano. Just don't blame Albert for all those reruns.

Internet and telephone communications are now transmitted primarily using fiber optic cables carrying our data in laser beams. The laser beams are detected at the receiving end by photoelectric devices. This advance allows the transmission of far more data, including pictures and videos, than is possible using copper wires. And we're saving money as well, because glass is cheaper than copper.

The photoelectric effect also lets us "see" things that our eyes can't. All space telescopes, including Hubble, and all major telescopes on the ground use CCDs to capture spectacular images of the cosmos. Everything we know about our universe, and our place within it, comes from observing light with photoelectric detectors. These devices are thousands of times more sensitive than our eyes or film, and they even detect light we can't see at all, such as infrared and ultraviolet.

Modern medicine has also advanced using the photoelectric effect. Instead of slicing patients open so doctors can look or reach inside, many examinations and surgical procedures are now done with ***endoscopes***, fiber optic cables equipped with photoelectric detectors. In abdominal surgery, for example, endoscopes and related tools are inserted through incisions that are less than a half-inch wide. These small incisions heal much faster and with far less pain than an 8-inch gash that surgeons need to insert both hands. As you lay spread out across the operating table, hoping the doctors really see what they need to, think of the man with the crazy white hair.

Aren't you glad Einstein got it right?

## SOLAR CELLS

Solar cells are based on the photovoltaic effect, which is very similar to the photoelectric effect. In both, an electric current is created by light hitting matter, but in somewhat different ways. In both effects, physicists had been unable to explain why low frequency light didn't produce the effect even at high intensity. Einstein's solution that light is both a particle and a wave explained both effects, and enabled development of effective and important products.

Solar cells are used in a variety of applications. Many watches and calculators are powered by light using the photovoltaic effect, thus eliminating the need for batteries. Solar cells are also beneficial for remote or mobile energy needs: powering lights on buoys; illuminating traffic signs; and providing electricity to cabins and campers. I've even seen solar energy collectors on the tops of mud huts in the Maasi Mara in Kenya. After a long, hot day of herding their cattle, even Maasi warriors want to put down their spears and relax with the latest sports news and reruns of Masterpiece Theatre.

Futurists are developing proof-of-principle cars and planes powered by solar cells. Solar-powered cars have been racing for over 20 years at a variety of venues, including Australia, North America, and South Africa. The Swiss electric airplane Solar Impulse completed a 26-hour, manned flight in July 2010. Its 12,000 photovoltaic solar cells powered a 10-horsepower electric motor and charged batteries that kept it aloft during 8 hours of darkness from sunset to sunrise.

The most anticipated application of solar cells is to replace fossil fuels in producing electricity for mass consumption. To achieve this, solar cells must become substantially cheaper and practical solutions must be developed to transport solar energy from the best collection sites, such as the Arizona desert, to consumers hundreds or thousands of miles away.

# 5

# Wavy Particles

At the core of Einstein's genius was his ability to see the underlying unity between things that outwardly seem entirely different. A prime example, discussed previously, was his discovery that light is ***both*** a wave and a particle. French physicist Louis de Broglie later generalized Einstein's idea by proclaiming that particles are also waves and have wavelengths. The resultant concept, called ***particle-wave duality***, has forever changed our view of the universe and everything in it.

## PHYSICS OF ATOMS

Niels Bohr, a Danish physicist and close friend of Einstein, used de Broglie's idea that particles have wavelengths to explain why atoms don't immediately collapse.

Wait—why ***would*** atoms collapse?

In the mid-1800's, British physicists Michael Faraday and James Clerk Maxwell developed Electromagnetism, a comprehensive theory of electricity and magnetism. This theory says that objects with electric charges emit radiation (light) when they are accelerated. Accelerations are any changes in an object's velocity—speeding up, slowing down, or turning (even if the speed doesn't change). For an object to go around in a circle, something must act on it and accelerate it. Earth orbits the Sun because it is accelerated by the Sun's gravity. The electric force between negatively charged electrons and positively charged nuclei accelerates electrons and holds them in atomic orbits. Electromagnetism says these accelerated electrons should emit radiation, lose energy, and fall closer to the nuclei, as shown in Figure 5.1. In fact, within a billionth of a second, all electrons should spiral into their nuclei, collapsing all atoms. Atoms would then be inert, unable to form molecules, making any type of life impossible.

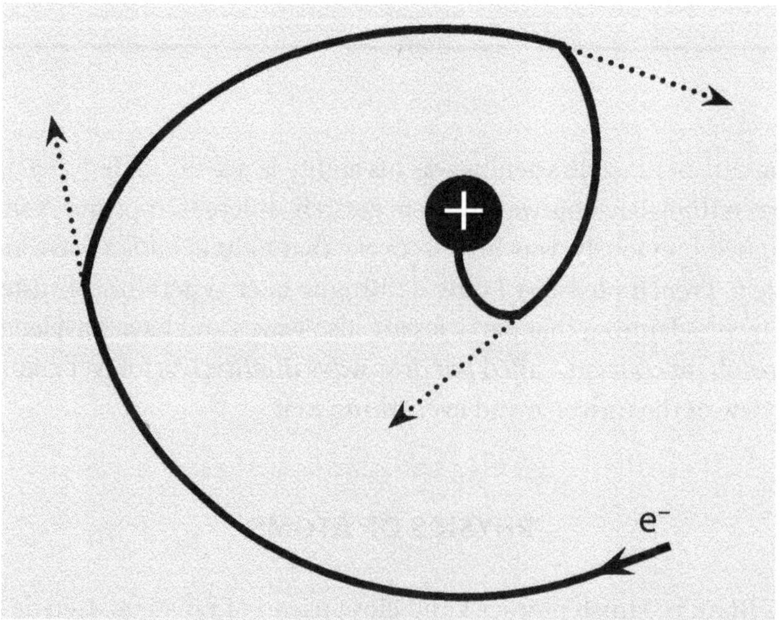


Figure 5.1. Before Quantum Mechanics, physicists thought an electron (e⁻, solid lines) orbiting a nucleus should lose energy by radiating photons (dotted lines) and spiral into the nucleus (central black circle. If so, all atoms would collapse, become inert, and life would be impossible.

Bohr showed that particle wavelengths prevent that catastrophe. Consider a single electron in a circular orbit around a nucleus, as shown in Figure 5.2. Label the circle like an analog clock face, with 12 o'clock on top and 6 o'clock at the bottom. At some moment, the electron's wave at

12 o'clock will attain its maximum height—it will crest. At that moment, consider the wave's height around the circle. Moving clockwise from the top, the wave height will go down and up and down and up as the wave oscillates. As we head back up to 12 o'clock, an interesting thing happens. Since waves are smooth, the wave's height at 11:59 has to match its height at 12:01—the wave must be at its maximum on both sides of the top. This means that in going around the orbit, the wave has to go down and up a whole number of times. The circumference of the orbit must be an integer number of wavelengths; it could be 1 wavelength or 7 wavelengths, but never 3.14 wavelengths. This is a unique property of the micro-world of atoms, and quite different from our everyday experience. By contrast, planets can orbit stars at any distance whatsoever—although we humans strongly prefer Earth's orbit, where temperatures are pleasant.

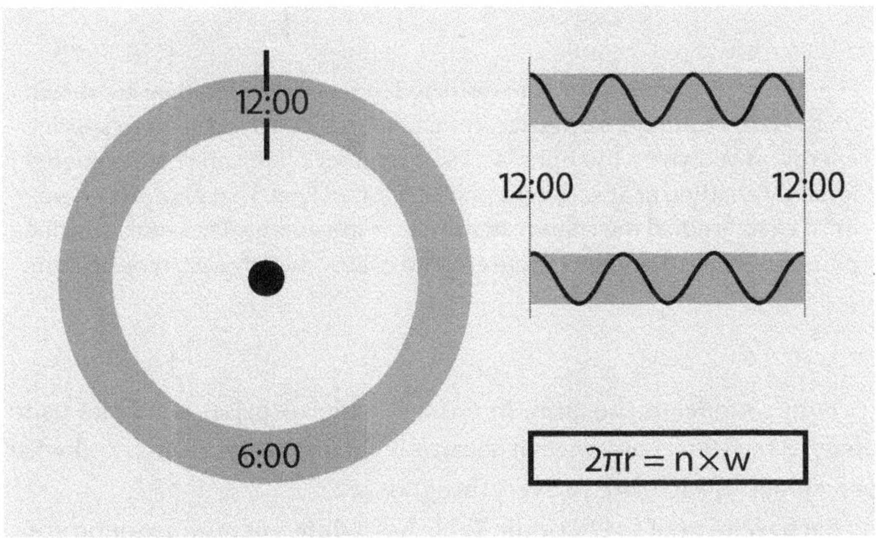


Figure 5.2. On the left, an electron orbits a nucleus (central black dot). To better illustrate our point, cut the orbit at 12:00 and straighten it as shown on the right. In the upper case, the electron's wave doesn't match at the two ends that join at 12:00. Bohr said allowed orbits must have electron waves that match on both sides of 12:00, as in the lower case. Thus, the orbit circumference ($2\pi r$) must be an integral multiple ($n$) of the electron's wavelength ($w$). The radius ($r$) of the electron's orbit can have only certain specific values.

Because electrons cannot have an orbit smaller than one wavelength, Bohr continued, they cannot spiral into the nuclei and atoms do not collapse—thank goodness.

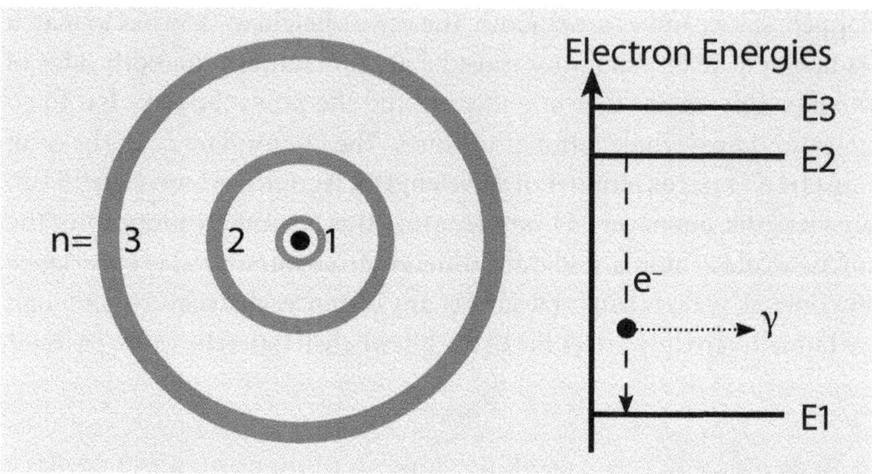


Figure 5.3. Electrons in atoms are restricted to specific orbits; three are shown on the left, orbiting a nucleus (central black dot). Each orbit has a specific energy, as shown on the right. An Electron (black dot) can change energy levels by emitting or absorbing a photon (dotted line). That photon's energy must exactly equal the change in the electron's energy. Thus, only specific photon energies (frequencies) are emitted or absorbed by each type of atom.

Bohr's model of the atom has another big surprise: it enables us to analyze the universe—we can ascertain what types of atoms, and what percentage of each, are in everything we see.

Each element of the Periodic Table has a different number of protons, and hence a different positive charge in its nucleus. Thus the electron energies allowed by the Bohr model are different from one element to another. When an electron drops from a higher energy orbit to a lower one, it emits light, as illustrated in Figure 5.3. Since the total amount of energy never changes (energy is conserved), the energy of this light exactly matches the energy lost by the electron. The opposite process also occurs—atoms can absorb light if the light's energy is exactly the right amount to raise an electron to a higher allowed orbit. Because each element has a unique set of allowed electron energies, each also has a unique set of light frequencies that it can emit or absorb. This unique set of frequencies, called a ***spectrum***, is like a fingerprint. By analyzing samples in their labs, scientists have cataloged the fingerprints of every element and many common molecules as well.

## ATOMS REVEAL WHAT'S INSIDE

In 1842, French philosopher Auguste Comte proclaimed: There are things mankind will never know. His prime example was the chemical composition of the stars. Comte seemed to be on solid ground picking stars. After all, how could we get to a star, how could we scoop up some star stuff, and how could we get it back to analyze in our labs on Earth? Yet, only a few decades later, Comte was proven wrong—at least regarding the composition of stars. As explained above, each type of atom emits and absorbs a unique set of light frequencies—a unique spectrum—that positively identifies it, just like a fingerprint, as Figure 5.4 illustrates. Analyzing the frequencies of light coming from any star, and comparing that with the known spectra of the elements of the Periodic Table, scientists can precisely determine how much of each element is in that

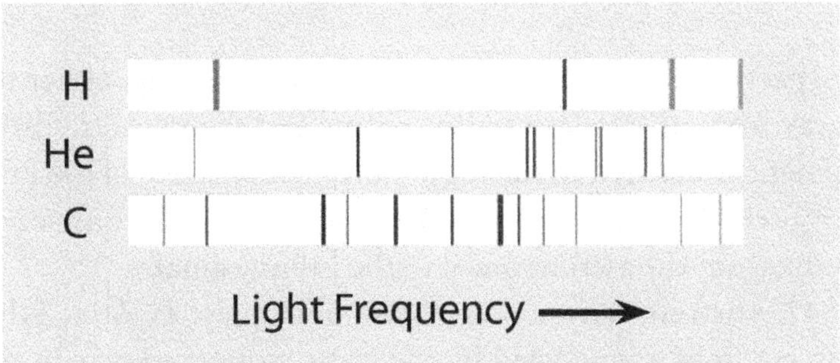


Figure 5.4. Above are the spectra of: hydrogen, H; helium, He; and carbon, C. The vertical lines indicate the frequencies emitted and absorbed by each element. Each type of atom and molecule has a unique spectrum, like a fingerprint, which identifies it anywhere in the universe.

star. They can easily do what Comte said was impossible, perform a chemical analysis of objects trillions of miles away, and still make it home for dinner. (Since these are astronomers, I should say: "and still make it home for breakfast.")

This process, called ***spectroscopy***, is now an essential tool in astronomy. It can tell us the precise chemical composition of every star, nebula, and galaxy in the universe, and it can even tell us how fast each of those bodies is moving (more on that later). Spectroscopy is so powerful that it allowed scientists to discover the element helium 93 million miles away from Earth before we ever found it in our own backyard. In 1868, French astronomer Jules Janssen observed a previously unidentified frequency in sunlight. Scientists eventually determined that this light came from a new element that they named Helium after the Greek sun god Helios. It was not until 1882 that Italian physicist Luigi Palmieri first discovered helium on Earth, in the lava of Mount Vesuvius.

Spectroscopy is also used closer to home to determine the composition of unknown substances. When CSIs want to identify trace evidence left at a crime scene, they often use a spectrometer. Samples are vaporized and their light spectra measured to determine the sample's atomic and molecular composition. Atoms tell all.

## ATOMS REVEAL MOTION

Atomic spectra also reveal how fast things are moving, either toward us or away from us. Understanding atoms and their spectra drove the development of modern cosmology, our understanding of the universe. This is a further demonstration of the intricate interconnections of everything from the astronomical to the infinitesimal.

In 1842, Austrian astronomer Christian Doppler explained that the colors in starlight are affected by the star's motion relative to us. The ***Doppler effect*** applies not only to light but also to sound, where it is more familiar. Most of us have heard the siren of an ambulance, police car, or fire truck as it drives by. As the siren approaches us, it has a high pitch, and as it passes, the pitch abruptly drops. Similarly, when a star moves toward us, we see the frequencies of its light increase, which we call ***blueshift*** as blue is the highest frequency visible light. Conversely, when a star moves away from us, we see its light frequencies decrease, which is ***redshift***. Because spectroscopy has become extraordinarily precise, astronomers are now able to measure a star's velocity, toward or away from us, to one-fifth of a mile per hour.

As a specific example of the Doppler effect, consider a star emitting yellow light. If the star weren't moving, we would see its light as yellow. If the star were moving toward us at 27 million mph, we would see a 4% higher frequency, which is green light, and if it were moving away from us at 27 million mph, we would see a 4% lower frequency, which is orange light. The frequency we observe changes by 1% per 6.7 million mph, even though the star is emitting exactly the same light.

American Astronomer Edwin Hubble used the Doppler effect to show that all distant galaxies are moving away from us—their light is redshifted. Furthermore, he found that the most distant galaxies are moving away the fastest, according to a simple rule we call ***Hubble's Law***. Hubble found the galaxies' velocities are proportional to their distances. Thus, if one galaxy is twice as far away as another, the first will be moving away twice as fast.

The simplest explanation of Hubble's Law is that our universe is expanding. And that means it must have had a beginning; it cannot have existed forever.

That's a lot to learn from a little atom.

## ATOMIC DATING

Another secret that atoms can reveal is an object's age. This is particularly remarkable since atoms are too small to have clocks. Atoms never age in the sense that all living things do. The hydrogen atoms in our bodies were formed over 13 billion years ago. These atoms are as pristine today as they were then, and they will probably remain pristine for at least another billion, trillion, trillion years. We may not be immortal, but it seems that our atoms are.

So, how can we use timeless atoms to tell time? While all atoms are timeless (no internal clock), some will last forever and some will not—the latter are ***radioactive***. How something without a clock can know when it's time to fall apart is one of the intriguing consequences of Quantum Mechanics and the waviness of particles. We'll save a full exploration of that mystery for chapter 11. But, to get a feeling for this, consider a bizarre analogy: imagine a group of lemmings intent on ending their lives according to a special ritual. Every morning at dawn, each lemming flips a coin and plunges off a cliff if his coin comes up tails. Odds are half of them will die each morning. If the group began with 1000 lemmings, there would be 500 the next day, 250 on the third day, 125 on the fourth day, and so on. This mathematical progression is an ***exponential decay*** with a ***half-life*** of one day. During any time interval equal to one half-life, the number of survivors is halved. These lemmings don't need a calendar to know when their time is up; that is determined by random chance and their half-life. Similarly, radioactive atoms don't need a clock; they have a certain chance of falling apart ("decaying") at each instant of time, according to the half-life of that type of atom. Half-lives vary tremendously, even among isotopes of the same element—100 billionths of a second for Thorium-218 versus 14 billion years for Thorium-232.

An important application of radioactivity is carbon-14 dating. Carbon-12 is the most common type of carbon and is not radioactive (as far as we know the half-life of carbon-12 is infinity). Cosmic rays hitting Earth's atmosphere create a small amount of carbon-14 that has a half-life of 5730 years. Carbon-14 is continually being created, and since it is radioactive, it is continually decaying. These competing effects achieve a balance—throughout Earth's biosphere, one in a trillion carbon atoms is radioactive carbon-14. Plants bring all types of carbon into the food chain. Since we are what we eat, one trillionth of the carbon atoms in all living things are carbon-14. Even though one in a trillion isn't much and 5730 years is a long time, we have so much carbon within us that 3000 carbon-14 atoms decay every second inside a 140-pound person. (Radioactive potassium-40 in our bodies produces over 4000 decays per second.) We are all slightly radioactive.

Only after death do we become less radioactive. Since carbon-14 atoms are no longer being ingested to replace those that decay, the abundance of carbon-14 decreases after death. After one half-life, 5730 years post-mortem, the level will be one in two trillion, and after ten half-lives, 57,300 years, the level will be one in a thousand trillion. Measuring the current level of carbon-14 in a dead organism determines when it died. Beyond ten half-lives, carbon-14 levels are generally too small for precise measurements.

To measure much greater ages, scientists use radioactive atoms with much longer half-lives. Uranium-238, with a half-life of 4.5 billion years, is a common choice for measuring the ages of meteorites and Earth's earliest rocks. Scientists can often obtain the ages of rocks that solidified billions of years ago to better than 1% precision.

# 6

# Beyond Black and White

As the last two chapters have described, Einstein broke down scientists' mental barrier that had long separated particles and waves. With de Broglie's extension, the resultant concept of particle-wave duality has dramatically changed our understanding of the universe.

We know now that "particle" and "wave" are labels, like "black" and "white", for opposite ends of a continuous spectrum, like the opposite ends of a ruler, as illustrated in Figure 6.1. Duality says that entities are not either at one end of the ruler or the other, but rather somewhere in between—everything in our universe is really a shade of gray. Something may seem more particle-like or more wave-like, but truly everything is a combination of both.

Figure 6.1. "Particle" and "wave", like "black" and "white", are labels for the opposite ends of a continuous spectrum. The reality of our universe is that everything is somewhere in between—a shade of gray, a combination of particle and wave properties.

Duality has profound and surprising philosophical implications. In the last chapter, we learned that duality underlies the stability of the atoms that make up everything we see, including our own bodies. But duality also clouds our future with impenetrable uncertainty.

## PHYSICS OF UNCERTAINTY

Physicists once thought particles, such as electrons, were miniature balls—like very small marbles. But, duality says even particles have wave properties, including wavelengths. This wavy nature of particles is the primary reason for the strange behavior of the quantum world with its inherent uncertainty.

Let's pause to define some terms used to describe waves. Frequency, f, is the number of wave crests per second that move past any point you care

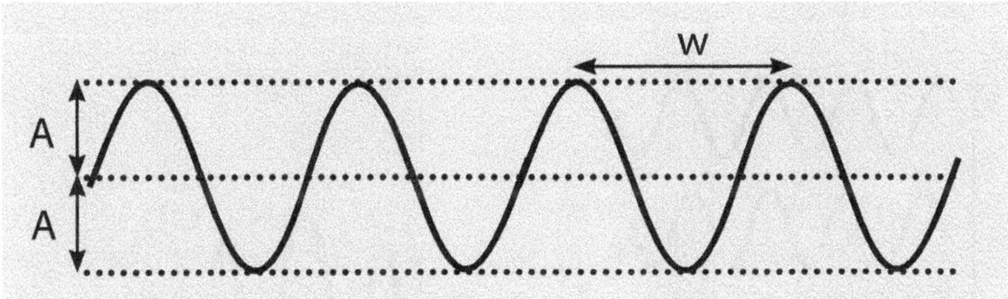


Figure 6.2. We characterize waves by their amplitude A (they oscillate between +A and −A) and their wavelength w (the distance between crests). The wave's speed, v, equals w times f.

to pick. Wavelength, w, is the distance between wave crests (or between troughs). Amplitude, A, is a wave's maximum value; the wave oscillates between +A and –A. See Figure 6.2. Wave speed, v, equals w times f. (It's always good to check an equation's units: for v=wf, the units are distance/second = distance/cycle × cycles/second, which is correct. An equation must be invalid if the units on both sides don't match.)

A pure wave, like a pure musical tone, oscillates at a single frequency and with the same amplitude everywhere. Thus, its energy is spread throughout space and the wave is completely un-localized. This is the opposite of what we expect from particles, which were always considered localized, existing at one definite, specific place. But if one adds together many waves of somewhat different frequencies, the sum is a ***wave packet*** that is partially localized; see Figure 6.3. Let's see why. Imagine summing waves of different frequencies that are all aligned to crest at one point called X. Since the waves are all at their maximum at X, their sum is very large. But at any other point, the waves will be out of synch because each has a different frequency and wavelength. Being out of synch, the waves will partially cancel one another. It turns out that the farther one moves away from X the more they cancel and the smaller the amplitude of the waves' sum. One can prove mathematically that the greater the range of frequencies in the wave packet, the more rapidly its amplitude drops, making the packet more localized. We call the spread in frequency df, and the spread in position dx. In fact, the math shows that df and dx are inversely proportional: df = (some number)/dx.

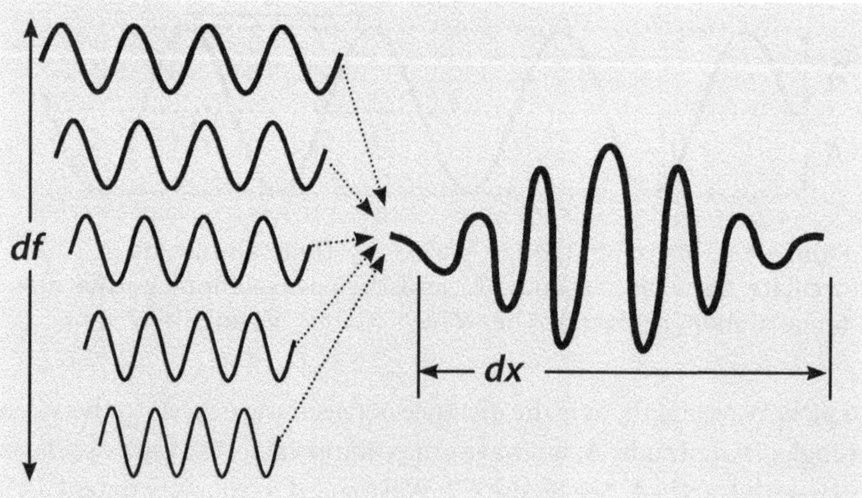


Figure 6.3. Pure waves have a single frequency and are spread evenly throughout space, as the five examples on the left. Summing waves over a range of frequencies, df, creates a wave packet, as on the right, which is partially localized in space, extending over a distance dx. To make dx smaller, one must make df larger, and *vice versa*.

A wave packet does not have one definite frequency, it contains a mixture of all the frequencies within df, hence its frequency is ***uncertain*** by the amount df. Similarly, the wave packet's location is ***uncertain*** by the amount dx. We can restate the equation of inverse proportionality to read: df × dx = (some number). Actually, this equation states the very best that is achievable; in less than ideal circumstances we might find that df × dx > (the same number).

De Broglie showed a particle's wavelength is inversely proportional to its momentum, p, which is related in a simple way to the particle's energy, E. Recalling v=wf, we also know that wavelength is inversely proportional to frequency, which is also related to energy. Putting all the equations together, one obtains Heisenberg's Uncertainty Principle:

$$dp \times dx \geq h/2\pi$$
$$dE \times dt \geq h/2\pi$$

Here, h is a number we call Planck's constant—h sets the scale at which quantum effects become important. In the first equation above, dp refers to the uncertainty in momentum in the x direction only; similar equations exist for y and z, the other two dimensions of space.

## DOES GOD PLAY WITH DICE?

The Heisenberg Uncertainty Principle says that for a particle, such as an electron, we can never know exactly ***where*** it is, and at the same time know exactly what its ***velocity*** is. As we discussed in the physics section above, we can reduce the uncertainty in location if we allow a greater uncertainty in velocity. Or, we can choose to reduce the velocity uncertainty if we allow a greater uncertainty in location. We can trade one for the other, but we can't improve both. We simply can't have everything we want.

This uncertainty seems strange. We have no trouble knowing the location of a baseball; why can't we know for sure where an electron is? Basically, it's because electrons are so much smaller. We know where baseballs are because light bounces off them and our eyes detect the reflected light. We see where the light comes from, so that must be where the baseball is. Why doesn't that work for electrons as well?

Let's give it a try. Imagine an electron orbiting the nucleus of an atom. Let's try finding where the electron "really" is. As with baseballs, we'll bounce light off the electron and detect the reflected light. The problem is that visible light has a wavelength that is much smaller than a baseball but much larger than an atom, a thousand times larger. Using visible light to peer into atoms is like tossing giant beach balls hoping to find a pea in total darkness. If a beach ball does bounce back, we'll know something is there but we won't learn much about what and where it is. No problem—we'll use higher energy light that has a higher frequency and a shorter wavelength. An x-ray with 5000 times the energy of visible light should do the trick because its wavelength is 5 times smaller than an atom. But now we have a new problem: the energy of the x-ray is 1000 times more than the energy needed to rip an electron away from its nucleus—our searchlight just became a sledgehammer. Yes, we can find out where the electron *was*, but in the process we'll blast it off into the wild blue yonder. By reducing the electron's location uncertainty we have drastically increased its velocity uncertainty. Heisenberg wins again.

We have no problem seeing baseballs because they're big. With visible light, we get a precise location without disturbance, because its wavelength is a million times smaller than a baseball while its energy is a billionth of a billionth of a "swat" by Babe Ruth. It's in the micro-world that Heisenberg reigns.

What does all this mean for us? At the atomic scale, we can never know enough to precisely predict the future. When particles interact they might bounce this way or that, or they might convert some of their energy into new particles. We might be able to learn that in a certain type of collision new particles are produced 40% of the time, but we will never be able to say for sure if that will happen in the next collision. This is similar to rolling dice. We know that rolling two die results in "snake eyes" (two 1's) once every 36 rolls, *on average*, but we can't predict on which roll snake eyes will come up next. In the macro-world, you might imagine precisely measuring each die, the properties of the table, and the motion of the shooter. In principle, if you knew everything you might be able to predict the outcome. In the micro-world, the outcome is unpredictable, even in principle, even if you know everything that is possible to know.

Einstein never accepted this view of nature, protesting: "God does not play with dice." He believed nature was simple, harmonious, and beautiful and therefore the true laws of physics should be elegant and perfect. He thought this theory of inherent uncertainty was ugly and imperfect—it just couldn't be complete, physicists must be missing something. Niels Bohr, leader of the opposing view, responded: "Don't tell God what to do with His dice." Bohr was convinced, and all the evidence has supported him, that uncertainty is an intrinsic property of nature, something we must accept whether we like it or not.

Einstein and Bohr (right) were great friends, who loved to discuss the most profound questions of existence, and frequently disagreed.

## PHILOSOPHY OF UNCERTAINTY

Duality tells us particles have wavelengths, and that in turns leads us inexorably to conclude that uncertainty is inherent in nature. Duality mandates an uncertain future. While we can still say some things are impossible, such as perpetual motion machines, we can never know for sure what the future holds. The best we can aspire to is determining the probability of specific outcomes—this horse is 37% likely to win the next race, which is still useful.

Thus, humanity must abandon Newton's notion, shared by Einstein, that the universe is like a giant clockworks, moving inexorably forward toward a precisely predetermined future. The two most famous physicists in history both believed that if we learned all nature's rules and measured exactly where everything is and what it's doing, we could precisely predict what would happen throughout the universe for all time to come. Newton and Einstein believed the future was predetermined, that there could never be any deviation from the preordained, and thus no free will.

Duality proves that world view is wrong. Precisely predicting the future is impossible—not just impractical, but completely and utterly impossible. Einstein hated that. But it resulted directly from the particle-wave duality that he himself had launched. He had no one to blame but himself.

Some scientists even believe that the uncertainty of the atomic world may have an important influence on the operation of the human brain—that some of what we call free will may actually be the end result of countless, random atomic events.

But they can't be certain.

# 7

# Lasers

The first operational lasers were built in 1960, five years after Einstein's death, but he published the fundamental theory that made lasers possible in 1917. At first, lasers were merely a scientific curiosity—people didn't develop lasers for any particular application, but rather to prove the science and show that the engineering could really be done. Once it worked, the laser was a solution in search of a problem.

Now lasers are nearly ubiquitous in our society—they sculpt our corneas, scan bar codes, read CDs and DVDs, carry telephone and Internet communications, and are the core technology in literally thousands of other applications. It might be easier to list the things that lasers are not used for.

Lasers work because particles of light, ***photons***, are gregarious—they "want" to be together in a common state; that is, they are drawn to move in the same direction with the same frequency. That seems like the friendly thing to do, but actually photons are the only common particles that do that. Particles of matter—protons, neutrons, and electrons—are fanatically antisocial. Under no circumstances whatsoever will two identical particles of matter ever share the same state; they each demand their own exclusive turf.

Because photons are gregarious, enormous numbers of them can be put into, and will remain in, a common state. This creates a powerful light beam that can travel enormous distances without losing its punch. With lasers, we can easily and economically deliver a precise amount of energy to a precise location, for a precise amount of time, and we can exactly reproduce that as often as we want for as long as we want.

Lasers are incomparable tools.

## PHYSICS OF LASERS

With Indian physicist Satyendra Nath Bose, Einstein showed that photons behave fundamentally differently than do particles of matter due to their ***spin***. Spin is a quantum mechanical property of all particles; it is a form of angular momentum that is intrinsic to each particle. Regardless of what else a particle does, it always has its spin. A useful starting point is to think of particles as little spinning tops, but as with all macro-world analogies, the micro-world of Quantum Mechanics is always different in some important way. Particle spins can only have certain specific values unlike macro-world tops that can spin at any rate. In appropriate units, the standard particles of matter (protons, neutrons, and electrons) each have spin ½ while photons have spin 1. Their different spin values are responsible for matter particles being antisocial and photons being gregarious. The antisocial behavior of spin ½ particles is described by Fermi-Dirac Statistics and these particles are called ***fermions***. The gregarious behavior of spin 1 particles is described by Bose-Einstein Statistics and these particles are called ***bosons***. More exotic particles have other spin values, but they also fall into the same two classes—those with integral spin values (0, 1, 2…) are bosons, and those with half-integral values (½, 1½, 2½…) are fermions.

How does spin make lasers possible? Recall that electrons must occupy specific orbits in atoms. We also know that everything in nature seeks the lowest possible energy state—that's simply due to probability. There are more ways for a rock to fall downhill and release energy than there are for that rock to absorb energy and move uphill. Electrons in higher energy orbits will eventually drop to a lower energy orbit (closer to the nucleus) if that lower orbit has a vacancy. When it drops, the electron will emit a photon whose energy exactly equals the energy that the electron loses, since the total amount of energy must be conserved. This is a quantum process with a certain probability per second of the electron dropping.

Now we get to Einstein's great idea. The probability of an electron dropping will be enhanced by the presence of other photons in exactly the same state as the photon that this electron will emit. For example, if an emitted photon enters the same quantum state as three other photons—bringing the total to four—the probability of the electron dropping will be 16 times greater (4 × 4) than if the three photons weren't present. This is illustrated in Figure 7.1. Einstein called this ***stimulated emission***, which ultimately led to the word "laser", an acronym for *l*ight *a*mplification by *s*timulated *e*mission of *r*adiation.

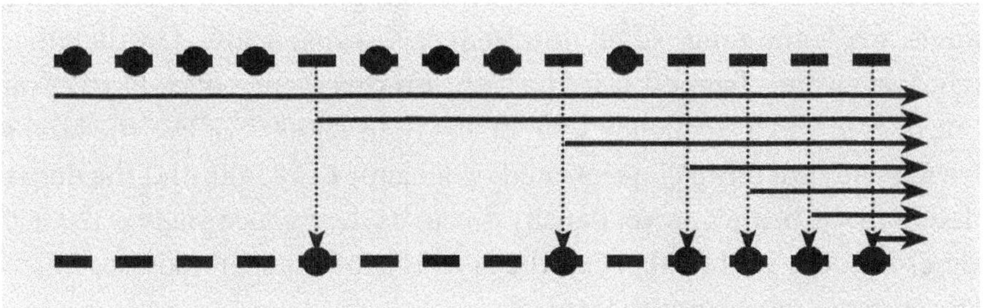

Figure 7.1. This image illustrates the creation of a laser beam that emerges at the right. Electrons (black dots) in many different atoms are "pumped" to a higher energy level (upper row of dashes). When a photon (upper solid line) of the correct energy passes, it stimulates electrons to drop to a lower energy level (bottom row of dashes) and emit matching photons. The greater the number of photons in the proper state the stronger the stimulation. Soon nearly all electrons drop and emit photons into a power beam—a laser beam.

Lasers are made using transparent materials in which the outer electrons can occupy two or more energy levels. Electrons orbiting many different atoms are moved to the higher energy levels ("pumping"). Since all the electrons are in the same type of atoms with the same energy levels, they are all primed to emit photons of the same energy (and frequency). When some of the electrons start dropping to the lower orbit, they emit photons that stimulate other electrons to emit matching photons, and very soon an avalanche results, producing a vast number of photons all moving in the same direction, all with the same frequency—a *coherent* light beam we call a laser.

## WHY LASERS ARE SPECIAL

Precision, control and power density are the prime attributes of lasers. While lasers come in a wide range of performance levels, let's compare a cheap laser to an ordinary flashlight. Flashlights work by heating filaments hot enough to glow and emit ordinary light that contains photons with a wide range of frequencies that move out in every direction. Even with mirrors and lenses, light from my flashlight diverges at a 28-degree angle, while the range of photon frequencies spans 50%. On the other hand, even the cheapest laser has a beam divergence of only $1/12^{th}$ of one degree and a frequency span of less than $1/1000^{th}$ of 1%. The laser's frequencies are more concentrated by a factor of 78,000, and the degradation of the beam's power density due to its divergence angle is 120,000 times less. Oh, and by the way, the laser is a lot cheaper and uses much less energy—what's not to love?

While most flashlight beams are worthless beyond several yards, laser beams remain remarkably focused over great distances, even from the Earth to the moon. In that case, astronomers use a better laser, one with a divergence angle of less than $1/1000^{th}$ of a degree.

Because a laser beam's photons are all traveling in the same direction, lasers are easily focused to minute spot sizes, providing outstanding resolution. Additionally, because laser beams are produced by electronics rather than by heat, they can be turned on and off ("pulsed") with superb control. Lasers pulses have been created with durations as short as 20 millionths of a trillionth of a second.

## WHAT LASERS DO FOR US

Many laser applications involve machining, the precise removal of material from an object or work piece. Laser beams can concentrate a large amount of energy in a very small area, and if necessary in a very short time. When a high-power laser beam hits a non-transparent object, atoms in the impact area immediately absorb the beam's energy and become so hot that they vaporize—the laser effectively drills a small cavity into the object. If the energy is delivered rapidly enough (high power for short duration), there won't be enough time for heat to flow beyond the impact area. Thus, almost all the laser's energy goes into vaporizing the target, minimizing damage to the rest of the work piece. By adjusting the beam size, energy, and duration, targeted material is removed with unparalleled precision. The beam can be electronically scanned across the work piece, removing unwanted material until it is fashioned as intended.

Laser machining can cut through the hardest metals or through the most delicate substances with great precision and minimal collateral damage. Laser engravings on fine woods are particularly appealing. Lasers are also used to restore priceless relics. Images of Saints Peter and Paul were recently discovered in ancient tombs in the catacombs of Rome, but were covered by 16 centuries of calcium carbonate deposits. Lasers were used to carefully remove deposits that were inches thick, without damaging the underlying paintings.

Laser cutting has enabled surgical procedures that would be far too risky with lesser tools. Lasers are used for many types of eye surgery, including LASIK for vision correction, retina reattachment, cataract removal, and glaucoma procedures. None of us want our eyes cut by a crude blade in an unsteady hand—computerized laser surgery is a great improvement.

Lasers have also improved heart surgery. Rather than opening a gaping hole in a patient's chest and connecting them to a heart bypass machine while surgeons operate, lasers allow many procedures to be done endoscopically. A small tube is inserted into the femoral artery of a patient's leg and maneuvered up to their heart. The target area is illuminated and imaged through fiber optics within the tube. Surgeons then send a laser beam through the glass fibers to remove life-threatening obstructions. Another advantage of laser surgery is that laser beams cauterize blood vessels (sealing them shut with heat) as they cut, thus reducing bleeding.

Lasers have almost limitless commercial applications. One of the first uses of lasers was bar code scanning at grocery stores, speeding up checkout and reducing errors.

The superior resolution achievable with lasers allows more information to be packed into a smaller space. Lasers are therefore used to read and write data, video, and music stored on CDs and DVDs. Higher frequencies have shorter wavelengths and thus support higher data densities, which is why companies advertize their "blu-ray" and not their "red-ray" devices.

Lasers have replaced many mechanical printers, both in commercial and consumer usage. Laser printers provide high-quality, high-speed, customizable printing, without the need for special paper. The printing process is xerographic and begins with an electrically charged drum. A laser beam scans across the drum, pulsing on and off, selectively neutralizing the drum's charge, and forming the desired image. Toner (dry ink) adheres to drum areas that remain charged, and is subsequently transferred to the paper. Initially, in 1981, laser printers had limited performance and cost $17,000, but now much superior devices sell for under $100—a fine example of how advancing technology can give us better products for less money.

Lasers are also critical in high-bandwidth communications, ever more important for telephone and Internet traffic. One hundred years ago, a one-minute cross-country phone call cost more than an average person earned in two days. Even in my youth, long distance calls were a luxury—we had to talk fast. Now we can send videos of our grandchildren to friends anywhere in the world at almost no cost. This advance is largely because communications are now carried by laser beams through fiber optics rather than by electrons flowing through copper wires. Old telephone wires carried 64,000 bits of information per second; lasers through fiber optics can carry 14 trillion bits per second, about 200 million times more. That spectacular bandwidth is sufficient to download the entire Library of Congress in less than one minute, but this isn't offered for residential use, at least not yet.

Want to know what time it is, ***exactly***? Ask a laser. Or rather ask NIST, the U.S. National Institute of Standards and Technology, to check their atomic clocks, each incorporating multiple lasers. NIST makes the world's most accurate timepieces, good to better than one part in $10^{17}$ or about one second per 4 billion years. If someone had started such a clock at the instant of the Big Bang, it would now be off by less than 4 seconds, 13.7 billion years later. With these atomic clocks, the U.S. Government will know exactly when it runs out of money.

Lasers can tell us where we are and how far we are from something else. Ring gyroscopes consist of a circular light tube with two laser beams racing around in opposite directions. Imagine the light tube is a donut mounted on an analog clock face. One laser beam circles clockwise while the other goes counter-clockwise. Since both lasers travel the same distance per loop, they take exactly the same amount of time. If both start off at the 12 o'clock position, they will always pass one another at 12 o'clock. Now imagine rotating the light tube clockwise. Because the velocity of light never changes and is unaffected by the motion of the light tube, the lasers will still pass one another where they did before. But now that point corresponds to a different time on the clock, because it was rotated. Let's say the laser beams now cross at 9 o'clock. From this change, we can determine precisely how much the light tube was rotated, 90 degrees in this case. If the light tube is attached to a ship, this means the ship turned 90 degrees to starboard. A ring gyroscope is much more accurate than any compass—better than 1% of one degree per hour. A device with a speedometer, three ring gyroscopes (each perpendicular to the other two), and a computer to keep track of everything, will always know how far and in what direction it has moved. This device is an ***inertial navigation*** system that doesn't rely on anything external, such as GPS, Earth's magnetic field, or the North Star. Such systems are used in aircraft, ships, and spacecraft, where laser-based systems have replaced mechanical gyroscopes.

Lasers are also used to measure distances. Since the speed of light has a well-known value that never changes, measuring the time required for a laser to travel from *here* to *there* and back determines the distance to *there*. Lasers measure distances small and large, everything from a billionth of an inch all the way to the distance to the moon. Special retroreflectors left on the moon by NASA astronauts make it possible to measure the lunar distance to about one inch, one part in 10 billion. These measurements show that the moon is moving away from Earth by about 15 inches per decade.

Laser distance measurements are being used to search for gravity waves—ripples in the fabric of spacetime predicted by Einstein to occur from calamitous cosmic events, such as the collision of two black holes trillions of miles away. A gravity wave hitting Earth might first stretch space, and everything in space, north-south while compressing it east-west. The expansion and compression directions can reverse thousands of times per second. The LIGO experiment is searching for these alternating distortions with ultra-precise laser distance measurements. They hope to measure distance changes to one part in 4000 trillion. They haven't found gravity waves yet, but the search continues. If they succeed, humanity will have a new type of telescope to view the heavens. Every time we introduced a new technology to examine nature, we discovered new phenomena. We don't know what we'll be able to "see" with gravity waves, but that's exactly why we want to do it—we'll never see if we don't look.

Astronomers use lasers to improve the image quality of ground-based telescopes. Ideally, all telescopes would be launched above Earth's atmosphere to avoid distortions caused by air turbulence and temperature changes, but this would be extremely expensive and inconvenient. Instead, lasers are aimed at the upper atmosphere where they excite atoms that emit light with specific frequencies—they effectively create fake stars. Telescopes on the ground detect the distortions in images of these fake stars and adjust their optics to compensate. This process can improve imaging resolution by a factor of 30, achieving resolutions as low as 30 milliarcseconds, about 1/100,000$^{th}$ of a degree.

While lasers can deliver enough power to vaporize steel, they can also cool materials down to extraordinarily low temperatures. The lowest possible temperature, the temperature at which there is no heat energy, is -459.67 °F, and is called ***absolute zero*** or 0 Kelvin. In some cases, lasers have cooled atoms down to 1/6000$^{th}$ of a degree above absolute zero. This technique is widely used by scientists, including in LIGO and in NIST for atomic clocks.

Tired of too much technology?

Just want something soothing to help you relax?

Go to a laser light show!

# 8

## Smarter Energy With $E=mc^2$

In 1905, Einstein made the shocking claim that mass was a form of energy. Why were people shocked? As with particles and waves, physicists had long been highly confident that energy and mass were two entirely different entities. They also were sure that energy was conserved and that mass was separately conserved—no reaction, they thought, could ever change the total amount of energy or change the total amount of mass. These beliefs were confirmed by numerous experiments, which by today's standards had limited precision and were restricted to low energy reactions.

Before Einstein, physicists knew energy came in many different forms, including: heat, work, light, electromagnetic energy, gravitational potential energy, chemical potential energy, and kinetic energy of motion. They also knew energy in one form could be converted, with some limitations, into energy in other forms—a candle converts chemical energy into heat and light.

Mass seemed entirely different. Every object we see has mass, which becomes clear if you try to lift it or push it. In chapter 3, we discovered that everything we see is composed of atoms, which are made of protons, neutrons, and electrons. Protons and neutrons have just about the same mass, but electrons are nearly 2000 times lighter. Thus, it's fair to say that any object's mass is determined by how many protons and neutrons it contains—mass is just a way of counting these nuclear particles. Weight is related to, but different from, mass. Weight is a combination of mass and gravity: an object's weight equals its mass times the acceleration of gravity at the object's location. Thus a hammer that weighs six pounds on Earth, would weigh one pound on the moon, and would weigh nothing inside the International Space Station. Yet, in each of these places, the hammer would have exactly the same mass.

While most forms of energy seem abstract—we can compute the amount of energy but can't directly observe it—mass is supremely tangible, nothing abstract about it. This is why Einstein's claim was so shocking. Once again, he saw the underlying unity between apparently disparate concepts and discovered that mass is really a condensed form of energy. We are all really made of energy. Furthermore, Einstein said that mass is not conserved, but can be converted into other forms of energy, such as heat, in accordance with his most famous equation, $E=mc^2$. Einstein said what is conserved is the total amount of energy in all its forms, including mass.

## WHAT DOES $E=mc^2$ REALLY MEAN?

This is the most famous of all equations—everyone has seen it, perhaps on a coffee mug, a T-shirt, or even in a physics book. But how many people would be comfortable explaining what it means? Actually, I believe that almost everyone understands the principle behind this equation. People understand this principle in a different context—the context of money. Not everyone likes numbers and equations, but we all pay more attention to numbers preceded by a $.

We all know that money, like energy, comes in many different forms—dollars, euros, pesos, and yen are all money—and we know that money can be converted from one form into another. If we fly from Los Angeles to Tokyo, we'll need to convert U.S. dollars into yen. The procedure is simple: the number of dollars we choose to convert multiplied by an exchange rate equals the number of yen we'll get. Einstein's equation says the same thing about mass and energy, as shown in Figure 8.1.

Mass can be converted into another form of energy, such as heat. The amount of mass being converted, that's the "m", times an exchange rate equals the amount of energy in the form of heat, that's the "E." The exchange rate is $c^2$, where c is the speed of light through empty space. That speed, c, is a very big number, 671 million miles per hour, and $c^2$ is c times c, which is a very, very big number.

Thus, Einstein made a profound statement: even a small amount of mass times this stupendous exchange rate yields an enormous amount of energy. In fact, the mass-energy in a single U.S. penny is enough to supply one million people with all their energy needs for a day. Equivalently, the mass of one penny has as much energy as we now get from burning two million gallons of gasoline.

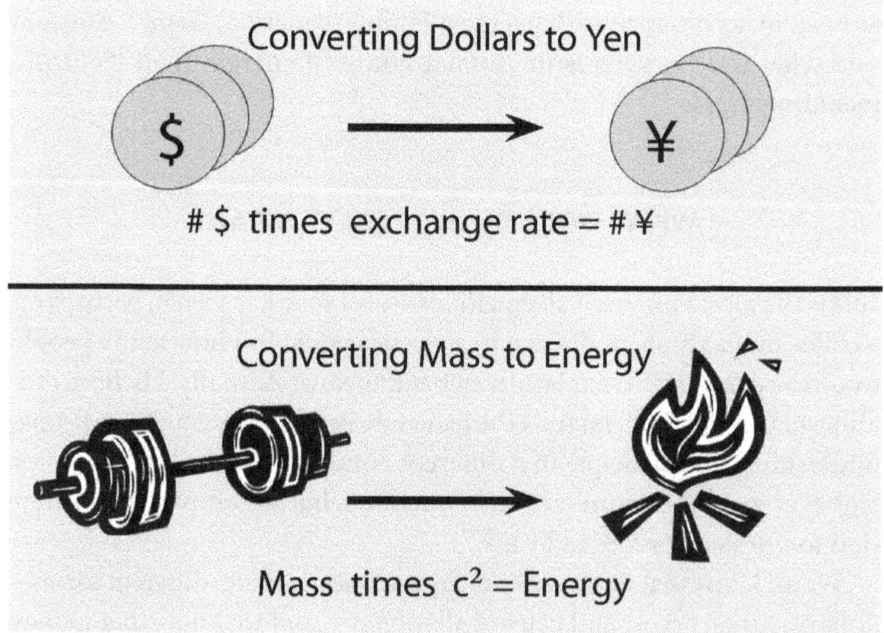


Figure 8.1. Money comes in many forms—dollars, euros, pesos, yen, and many more. Money in one form can be converted into another, as at the top: the amount in dollars times an exchange rate equals the amount in yen. Similarly, mass can be converted into other forms of energy, such as heat, as shown below. The amount of mass, m, times an exchange rate, $c^2$, equals the amount of heat energy, E.

## SMARTER ENERGY PRODUCTION

Let's use Einstein's equation, $E=mc^2$, to address a critical challenge we all face: how are we going to get all the energy our society needs and wants without bankrupting our economy and destroying our environment?

Everyone knows energy is now very expensive. The U.S. spends about $500 billion each year on foreign oil—money we dearly need at home. Our dependence on foreign oil has enormous geopolitical implications that often force us into situations that we would much rather avoid. And burning all that fossil fuel produces an enormous amount of pollution, including toxic chemicals that are poisoning our air, water, and soil. We simply cannot afford to continue producing energy with 19$^{th}$ century technology.

Fortunately, Einstein and nature can help us find ways to do much better. In a recent year, the total amount of energy that Americans consumed, from all energy sources, was equivalent to 25 trillion kWh. That turns out to equal the energy in one ton of mass. Thus, the challenge we really face, and the best way to view it, is how to best convert one ton of mass into useful energy each year. What are our options?

**Options to Provide U.S. Annual Energy Usage**

| Option | Tons of Fuel Required | Clean? |
|---|---|---|
| Burn Coal | 5,000,000,000 | NO |
| Burn Gasoline | 2,000,000,000 | NO |
| Nuclear Fission | 50,000 | no |
| Nuclear Fusion | 133 | yes |

If the U.S. produced all its energy by burning coal, it would need to burn 5 billion tons of coal each year. Only one ton of those 5 billion tons would be converted into energy and all the remainder would be converted into pollution—horrendously inefficient and dirty.

Gasoline has more pop per pound than coal. Converting one ton of mass into energy would require burning 2 billion tons of gasoline. We would again be left with billions of tons of pollution. Better than coal, but still awful. And since petroleum is an expensive and limited resource, we must find a better alternative.

Why can't we get more energy from these fuels? If we could be three times more efficient, we would need three times less fuel, energy would cost three times less, and we would generate three times less pollution. Wouldn't that be great?

The reason we can't do much better with chemical fuels lies within the atom. All fuels are made of atoms—everything is—so producing energy from any fuel boils down to getting energy out of its atoms. As discussed in chapter 3, atoms are simple—they have only two parts that we need to consider here: a nucleus surrounded by a cloud of electrons. The nucleus typically has 4000 times more mass than the electrons, and can have millions of times more extractable energy. When we burn coal, gas, or any other chemical fuel, what happens fundamentally is that the atoms change the way they share their electrons but the nuclei are unchanged. The electrons release some of their energy, but they only have a sliver of the total energy. Since the nuclei are unchanged, none of their vastly greater energy is released; see Figure 8.2. As long as we're only rearranging electrons, we will be grossly inefficient, continue to have expensive energy, and continue to produce enormous amounts of pollution.

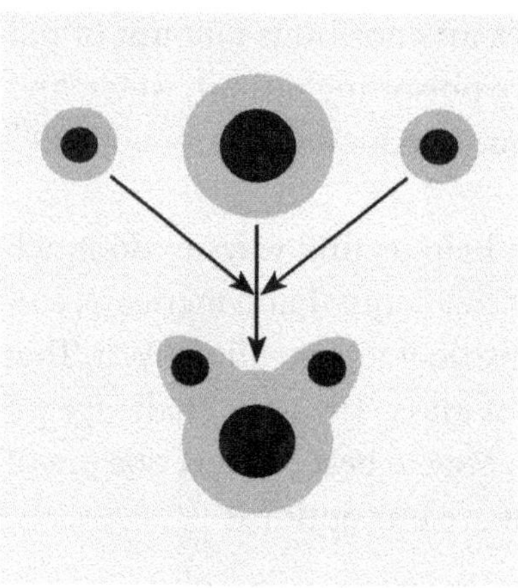

Figure 8.2. Two hydrogen atoms and one oxygen atom, above, react to form one water molecule, seen below. In any chemical reaction, the nuclei are unchanged and release none of their enormous energy. Only the electrons change states, releasing some of their much smaller energy.

If we want to have much cheaper and cleaner energy, we have to go where the energy is in atoms—in the nucleus. There are two ways to get energy out of nuclei: make them bigger, or make them smaller.

Let's discuss each.

Making a nucleus smaller is called nuclear fission; it's the process in operation in nuclear power plants today, and is illustrated in Figure 8.3. In nuclear fission, a very large nucleus, such as uranium-235, splits into two still quite large nuclei. The good news here is that fission is 100,000 times more efficient than burning coal. The bad news is that the end products of nuclear fission are radioactive. If the U.S. produced all its energy from uranium fission, that would require 50,000 tons of uraniumn-235 annually. One ton would be converted into energy and we would be left with 49,999 tons of radioactive waste. The trouble is none of us want that waste buried in our backyard.

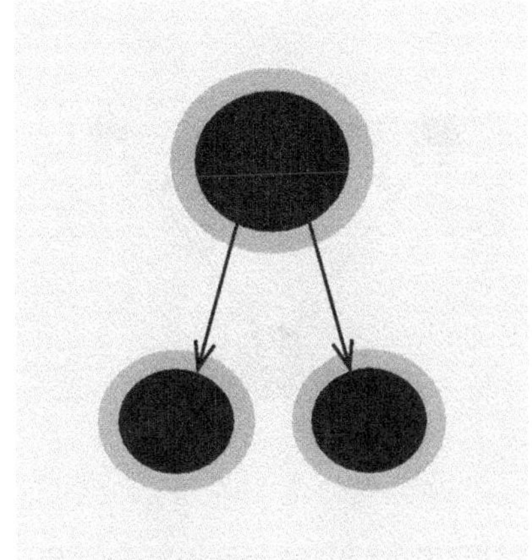

Figure 8.3. In nuclear fission, a very large nucleus like uranium splits into two or more smaller nuclei. Fission typically releases 100,000 times more energy than chemical reactions, but produces radioactive end products.

Making nuclei bigger is called nuclear fusion, and is the opposite of fission in nearly every respect. Fusion merges small nuclei, such as hydrogen, to make larger nuclei, such as helium, as illustrated in Figure 8.4. The good news here is that fusion is ***40 million*** times more efficient than burning coal. The other good news is that there is no radioactive waste. The end products of fusion are helium, carbon, oxygen, nitrogen, iron, and other similar elements. Nuclear fusion produces the atoms that make life possible—every carbon and oxygen atom in the entire universe was produced by nuclear fusion. This is the way nature makes energy—nature does not pollute, nature is not inefficient, nature has been making energy the right way for billions of years. Every star in the universe is powered by nuclear fusion. Nearly 90% of the atoms in our bodies, and over 99.99% of Earth's atoms, were created in the center of stars by nuclear fusion, and returned to the cosmos by stars' explosive deaths.

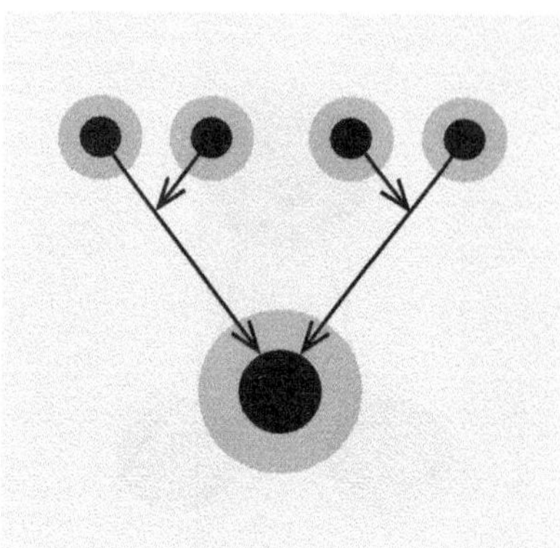

Figure 8.4. In nuclear fusion, two or more smaller nuclei merge to create a larger nucleus. Fusion can release 40 million times more energy than chemical reactions, and produces no radioactive waste. This is the way nature makes energy—efficiently and cleanly.

Gas clouds enriched with those atoms later collapsed to form new generations of stars, planets, and ultimately us. We are indeed made of stardust—the twinkle in your loved one's eye really is a bit of a star. The U.S. could get all its energy from the fusion of 133 tons of hydrogen each year. One ton would become energy and the other 132 tons would become helium, a perfectly clean substance that we could use to fill party balloons and blimps. If we vented helium into the air, it would float to the top of the atmosphere and go off into outer space. Since 24% of the universe is made of helium, that wouldn't even pollute outer space.

Scientists have been working on controlled nuclear fusion for 50 years, with only limited progress. But in the last few years, a new approach has been proposed that is vastly simpler and may make nuclear reactions similar to fusion a commercial reality. This new concept, called LENR (Low Energy Neutron Reactions), bypasses the greatest obstacle to fusion: how to squeeze together two positively charged nuclei. Fusion scientists have not yet solved the basic challenge that positive electric charges, like two nuclei, strongly repel one another. The LENR approach is to make neutrons, which have no electric charge, and use them to initiate nuclear reactions similar to fusion. The energy production efficiency of LENR should be comparable to fusion, and much easier to achieve. LENR should produce no radioactive waste, no chemical pollution, and no greenhouse gases. LENR devices could be much more affordable and easily portable. If LENR can be made a practical reality, energy will become very cheap, pollution-free, and abundant. This will be the greatest advance in human civilization since we mastered fire.

## NUCLEAR WEAPONS

Einstein's least favorite legacy was certainly nuclear weapons. For much of his life, Einstein was an ardent pacifist who actively encouraged young men to refuse military service. The rise of Nazism convinced Einstein to grudgingly accept that democracies needed armies to prevent despots from conquering the world, and that war was sometimes necessary.

In 1939, Einstein wrote to President Franklin Roosevelt informing him that the Nazis were developing an atomic bomb that could be devastating. He urged Roosevelt to initiate an American effort to counter this threat. Roosevelt launched the Manhattan Project, which succeeded in producing the first nuclear weapons, two of which were dropped on Hiroshima and Nagasaki in August 1945, ending World War II.

Those two bombs, quite small compared with today's weapons, killed an estimated 200,000 Japanese civilians; some died immediately from the blast and some died later from radiation exposure. Debate continues to this day about the morality of using such weapons.

Undeniably, every major combatant in World War II employed terror bombing—the deliberate targeting of civilians—with the goal of breaking their enemies' will to fight. With conventional weapons, that strategy never succeeded, despite killing millions of civilians. Nuclear weapons brought devastation to a previously unimaginable level of horror—that was the intent, and that is why it succeeded.

When President Harry Truman made the decision to drop atomic bombs, the Battle of Okinawa had just ended. It was one of the bloodiest battles of a horrific war; the U.S. suffered 12,000 deaths and 48,000 casualties. Japanese deaths included 127,000 troops and 100,000 civilians, one quarter of the island's civilian population. Everyone expected even fiercer resistance if the U.S. invaded Japan's home islands, where there were 300 times as many Japanese troops and 180 times as many civilians. Simply scaling by these numbers yields mainland invasion fatality estimates of several million Americans and many tens of millions of Japanese.

The U.S. Joint Chiefs of Staff estimated that invading Japan would result in 1.2 million U.S. casualties with 267,000 deaths, but U.S. military estimates were typically optimistic. The State Department estimated up to 4 million U.S. casualties with 800,000 deaths, and 5 to 10 million Japanese deaths. In anticipation of the invasion, the U.S. Army procured 500,000 Purple Hearts. As that invasion never happened, they still have 120,000 of those medals in stock after 60 years and several other major wars.

While the calculus of war is horrific, there is an important difference between hundreds of thousands of deaths and tens of millions of deaths. Certainly dropping atomic bombs created a catastrophe, but it was no doubt a lesser one than that of an invasion.

# 9

# Special Relativity

Einstein's theory of Special Relativity changed everything physicists believed about space, time, and motion. It superseded Newton's Laws of Motion, which had been the gold standard of science for 250 years. Never has a scientific revolution been so sudden and so comprehensive, either before or since.

Einstein based his theory on two assumptions: the Principle of Relativity and the constancy of the speed of light. He didn't prove these assumptions were correct—he assumed they were true, developed a complete theory based on them, and predicted what would happen if his theory were correct. In the subsequent 100 years, tens of thousands of experiments have tested his predictions and Special Relativity has never been proven wrong. Indeed, Special Relativity has been more thoroughly and precisely tested and validated—to as many as 18 decimals places—than has any other concept of science or other field of human endeavor. Nothing mankind knows about the physical world is more certain than is Special Relativity.

## MOTION IS RELATIVE

What do physicists mean by "relativity"?

Relativity is an old idea; Galileo first proposed it as a principle of science in 1632, and it was incorporated into Newton's Laws of Motion. Just as mankind learned that Earth is not the center of the universe, the Principle of Relativity states there is no privileged vantage point that nature prefers to all others. Thus, we can choose any vantage point, any ***reference frame***, to observe or measure any natural process and we will always find the same laws of nature. Special Relativity and Newton's Laws require only that the reference frame move with a constant velocity, which we'll assume for the rest of this chapter. On that basis, the laws of nature are equally valid in all reference frames. If a preferred reference frame did exist, we could definitively say: "this jet is flying 600 mph"—the jet's velocity would be ***absolute***. But, since all frames are equally valid, we must be more specific and say: "this jet is flying 600 mph ***relative*** to the ground", or relative to that aircraft carrier, or relative to the air.

Different reference frames are more useful in different circumstances. Velocity relative to an aircraft carrier is particularly important for a jet about to land. Someone might object to using an aircraft carrier as a reference frame because it moves. But actually, everything moves. Air moves with the wind. The ground moves as Earth rotates once a day. Earth orbits the sun at 70,000 mph. Our Sun orbits the galaxy at 500,000 mph, and our galaxy is moving toward Andromeda. There is no "stationary" reference frame.

Fortunately, the Principle of Relativity says we don't need a stationary reference frame because absolute velocities are not physically meaningful, only velocities measured relative to something else are meaningful. Consider an example: imagine two laboratories, each containing a scientist and lots of instruments, as shown in Figure 9.1. Now imagine completely sealing those laboratories from the outside world and placing each in a large cargo plane. The planes fly off in different directions at different speeds. The Principle of Relativity states that any experiment the scientists perform will yield exactly the same results in both planes, regardless of their speeds or directions—there is no way for either scientist to

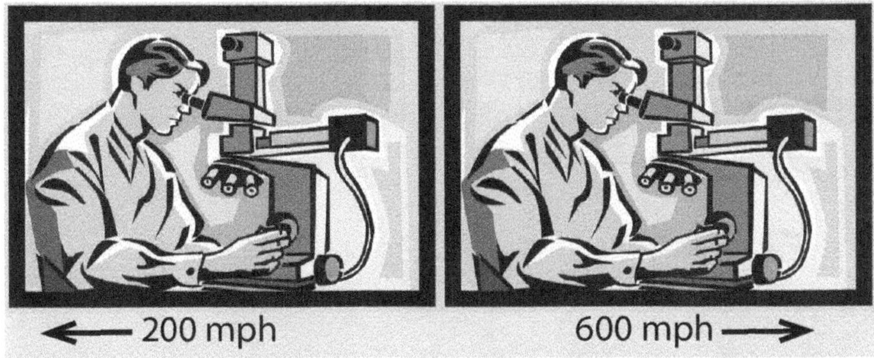


Figure 9.1. Two scientists in sealed labs cannot determine how fast their labs are moving. Any experiment they perform will yield exactly the same result, regardless of their velocities. The Principle of Relativity states that only the relative velocity between two objects can matter.

know the velocity of his plane without being able to reference anything outside their labs. Of course, one could drill a hole through the seal and stick out a GPS antenna, but that's cheating.

The Principle of Relativity was accepted for over two centuries. But in the mid 19th century, Maxwell and Faraday developed the Theory of Electromagnetism ("EM"), which seemed to require an absolute frame of reference. It seemed that either: the Principle of Relativity was wrong or EM was wrong—one would have to go. Because EM was so successful in so many important applications, almost every physicist was prepared to abandon relativity. Einstein believed both were correct, and he started searching for a way to make them compatible.

## THE SPEED OF LIGHT IS CONSTANT

Einstein discovered that EM and relativity could both be true only if light has a unique property, something that no one else had ever considered possible. Light, Einstein said, is a wave that does not require a medium (such as ether) to travel thorough, as sound waves require air. He said physicists had failed to find the luminiferous ether because it simply does not exist. He said light travels through empty space at a constant

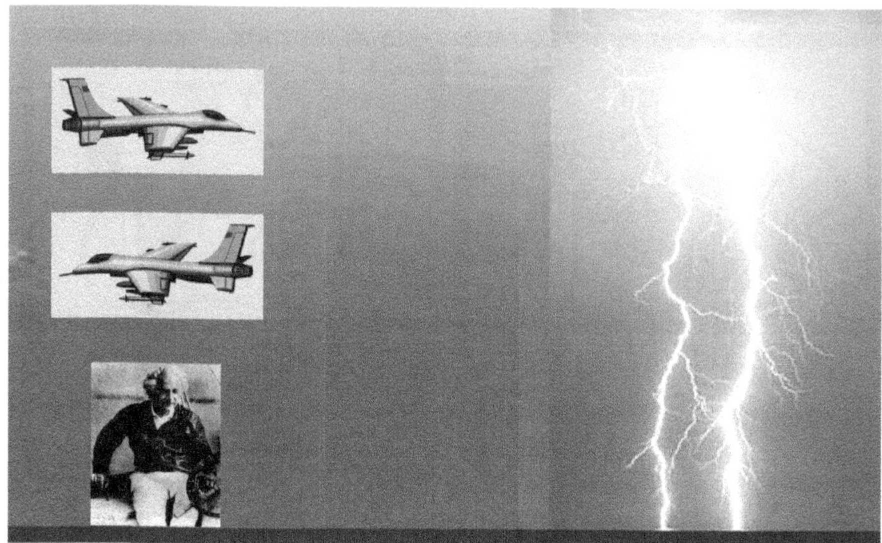

Figure 9.2. Einstein and the crews of two jets observe lightning. Each observes sound reaching them at different velocities due to their different states of motion. But, each will observe light reaching them at exactly the same velocity, c.

speed under all conditions, a speed we call c. Light's speed is lower in glass or other materials, but to simplify our discussion, in this book, "the speed of light" will always refer to c, light's speed through empty space. Einstein said regardless of the speed of the light source, or the speed of the light detector, or the motion of the observer, everyone will always measure the same value for the speed of light, c.

Why is that unusual? Consider Figure 9.2. A spectacular lightning strike is observed by Einstein and by the crews of two jets. While Einstein is calmly sitting still, the jets are flying 600 mph, with the higher jet flying toward the lightning and the lower jet flying away from it. Einstein observes the lightning's sound coming toward him at 768 mph, the speed of sound through air under normal conditions. But because the jets are moving through the air, the crew of the upper jet observes the lightning's sound coming toward them at 1368 mph (768+600), and the crew of the lower jet observes the sound approaching at 168 mph (768–600). This is the way velocities normally combine, and what we intuitively expect. But light, Einstein said, behaves very differently—he and the crews of both jets observe light from the lightning strike traveling at exactly the same speed, c, regardless of the observers' very different velocities. That *is* unusual.

## WHAT'S RELATIVE AND WHAT ISN'T

Occasionally, at cocktail parties and such, someone declares: "It's OK, Einstein said ***everything is relative.***" Often that's an excuse for aberrant social behavior. Actually, that's not what Einstein said. He said the laws of nature are universal—the same for all observers in all reference frames—nothing relative about that. It's certain measurements that are relative.

Imagine measuring the length, weight, and clock rate of two identical jets, one on the tarmac and the other flying past at 0.9c, 90% of the speed of light. Einstein said the lengths, weights, and clock rates of the two jets will appear different, because these measurements are relative, dependent on the relative motion of the observer and the observed. But if we plug these relative values into the equations of physics, the differences will cancel out—all the laws of nature will apply identically, regardless of the reference frame's state of motion.

Einstein said a moving jet will appear to us to be shorter, heavier, and its clock will run slower, all changed by the same factor, as compared with a stationary jet. Furthermore, he said, if the jet's crew looks back at the jet on the tarmac, they will say that jet is shorter, heavier, and its clock is running slower than their jet. This is because motion is relative; no one can say who's really moving and who's not; all we can agree upon is how fast we are moving relative to one another.

## SPEED SLOWS TIME

One of the most surprising deductions of Special Relativity is its new concept of ***time.*** Newton believed time was absolute and universal. He thought time flowed at its own intrinsic rate unaffected by anything else, and flowed at the same rate everywhere and always. A single clock, perhaps Big Ben, would be sufficient to keep track of Universal Time. Anyone could phone 1-800-4BIGBEN and a lovely voice would say: "At the beep, Time everywhere in the universe will be..." Einstein showed all that was wrong. Time is relative; there is no one "true" time. Time runs at different rates in different reference frames; different observers measure different times.

Let's examine why this is true. Time is our way of measuring how rapidly things change. If nothing ever changed, time would be meaningless. Every clock counts the number of "times" something changes.

So, let's try a ***thought experiment.*** Einstein loved thought experiments; these were the only experiments he ever did. In thought experiments, Einstein could keep his hands clean and avoid high voltage shocks, liquid nitrogen burns, and ionizing radiation (all the things that made me the man I am today). Thought experiments eliminate as many practical limitations as possible and focus on key principles. As Einstein said: "Make it as simple as possible, but not *too* simple."

So here we go: imagine an ideal clock made of two perfectly reflective, exactly parallel mirrors with a single photon bouncing back and forth between them, as shown on the left side of Figure 9.3. When the photon hits the top mirror, our clock counts one tick. Now add a second identical clock moving to the right at 90% of the speed of light, in the jet discussed above. The jet's crew would see their clock as stationary and its photon going straight up and down. But we see everything in the jet moving rapidly, including the clock's mirrors. Thus, the photon in the jet's clock appears to us to have a longer path to travel to register one tick. Remember, both photons must travel at the same speed, c. While our stationary clock counts 100 ticks, we see the moving clock count only 43 ticks. We observe the moving clock measure time passing more slowly than does our stationary clock.

Should the jet's crew get a better clock? That won't help. The Principle of Relativity requires all clocks moving at the same velocity to measure time at the same slower rate. If a second type of clock measured the "correct" time, the jet's crew could use the difference between these two types of clocks to determine their absolute velocity, which the Principle of Relativity says is impossible.

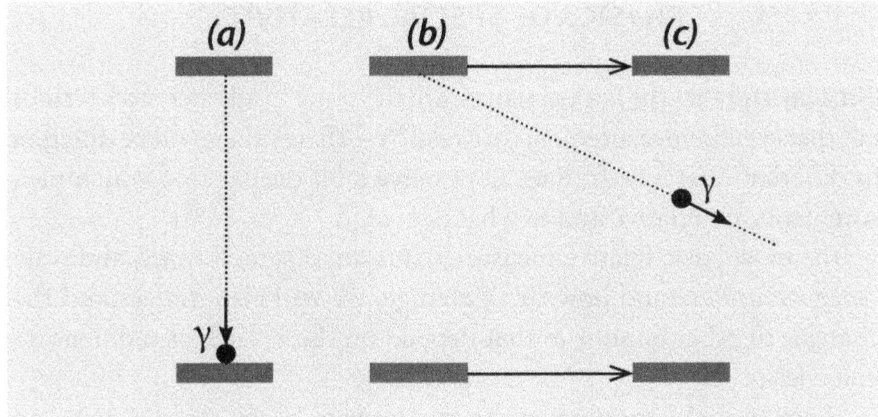


Figure 9.3. A stationary clock on the left has two mirrors and a photon, γ, bouncing between them; it counts one tick each time the photon hits the upper mirror. A second identical clock moves to the right at 90% of the speed of light. Both photons move at the same speed, but the one on the right appears to us to have a longer path to travel between mirrors. While we see the stationary clock count 100 ticks, we see the moving clock count only 43 ticks. We see everything in the moving frame change slower; we see ***time*** itself running slower.

Everything in the jet's moving frame seems to us to be running slower, not just clocks. We observe the crew's hearts beating slower, their brain waves waving slower, and their hair growing slower. Every biochemical process in their bodies appears to proceed slower, their lives proceed slower and last longer, and all at exactly the same slower rate. In truth, it is ***time*** itself that appears to us to be slower in a moving frame; this is called ***time dilation***.

Relativistic effects are symmetric. The jet's crew would look back and say their clock is fine, but our clock is running slowly. They would say we are shorter and heavier. They would observe all the same strange effects in our frame that we observe in theirs. Who is right? ***We both are!*** This seems impossible, but it all boils down to how measurements are really made. Since the speed of light is not infinite, moving people and stationary people simply do not make the same measurements. There is no single correct answer to "what time is it?" The answer really is different in different reference frames.

## PHYSICS OF SPECIAL RELATIVITY

Einstein said that the laws of nature are the same in all references frames, but that certain measurements are relative—their values will be different in different reference frames. Let's delve a bit deeper into which measurements are relative and to what degree.

The most basic relative measurements are distance, mass, and time. Once we understand how these change, we will also understand the changes in other quantities that depend on these, such as momentum and energy.

Einstein said a moving jet appears to us to be shorter, heavier, and have a slower clock than an identical jet that is not moving relative to us—each changing by the same factor *g*, as shown below.

| Key Measurements that are Relative | | |
|---|---|---|
| Measurement | Parked Jet | Moving Jet |
| Length | d | d / *g* |
| Mass | m | m × *g* |
| Time Interval | t | t / *g* |

The moving jet appears shorter only along its direction of motion, call that its length; its height and width do not change. (Physicists typically use $\gamma$ as the symbol for this relativistic factor; in this book, we use *g* to avoid confusion with the symbol for photons.)

We can use the clocks of Figure 9.3 and simple geometry to derive the formula for *g*. The photon in the moving clock must traverse the hypotenuse of a right triangle to reach the lower mirror. The vertical side of that triangle has the same length as the path traversed by the photon in the stationary clock; that length is ct, where c is the speed of light and t is the time required to reach the lower mirror in the stationary clock. In our reference frame, the photon in the moving clock takes *g* times longer to reach its lower mirror, thus its transit time is t*g* and the length of the hypotenuse is ct*g*. The length of the horizontal side of the triangle is the distance the jet has moved during time t*g*, which equals vtg, where v is the velocity of the jet. Pythagoras said $(ctg)^2 = (ct)^2 + (vtg)^2$. Rearranging and dividing each term by $c^2t^2$, yields $g^2(1–v^2/c^2) = 1$. The table below lists values of *g* for various velocities.

| Relativistic *g* factor vs. Velocity | | |
|---|---|---|
| **Velocity vs. c** | **Velocity in mph** | ***g* factor** |
| 0 | 0 | 1 |
| 0.0015 | 1 million | 1.000,001 |
| 0.15 | 100 million | 1.01 |
| 0.5 | 336 million | 1.15 |
| 0.8 | 537 million | 1.67 |
| 0.9 | 604 million | 2.3 |
| 0.99 | 664 million | 7.1 |
| 0.999 | 670 million | 22.4 |
| 0.999,999,999,7 | 671 million | 40,000 |
| 1.0 | 671 million | Infinity |

Notice that *g* increases rapidly for velocities above 0.9c. The fastest charged particles that I've worked with had v = 0.999,999,999,999,7c, and *g* = 40,000. Note also that for v=c, *g* would be infinity, which tells us something is wrong. For any object with nonzero mass, we would have to supply an infinite amount of energy to accelerate it to v=c. Since there isn't an infinite amount of energy in the observable universe, reaching or exceeding the speed of light is impossible for any object with nonzero mass. Conversely, photons, which have zero mass, must always move at v=c (in a vacuum) or else their energy would be zero and they would not exist.

Before Einstein, no one had observed these changes because they are extremely modest at normal velocities, velocities much less than the speed of light, where the value of *g* is extremely close to 1. For example, at 1000 mph, these changes are only one part in a trillion (*g* = 1 plus one trillionth). Appendix B explores a remarkable situation in which the extremely modest effects of relativity at small velocities accumulate to produce spectacular consequences. Nonetheless, in the vast majority of situations, relativistic effects are important only at velocities approaching the speed of light.

## EINSTEIN AND SPACE TRAVEL

Special Relativity has dramatic impacts on our hopes for future space travel, both favorable and unfavorable. The favorable impact is that astronauts will age slower and live longer in super high-speed rockets. The bad news is that they will never be able to go fast enough to return before everyone they knew on Earth had long since died.

With current technology, exploring beyond our solar system is completely unrealistic. The nearest star to the Sun, Proxima Centauri, is 4.2 ***light-years*** away. Since one light-year is the distance that light travels in one year, nearly 6 trillion miles, Proxima is 25 trillion miles away. A round trip to Proxima in the fastest rocket made so far would take over 100,000 years—just imagine how many times someone might ask: "Are we there yet?" But why would we go there? We already know there's no suitable planet orbiting Proxima. I estimate habitable planets are so rare that the nearest one might be 300 to 3000 light-years away, making a round trip to the next oasis a multi-million year odyssey, with current rockets. The cost of space travel is also a staggering challenge; one Space Shuttle launch cost $1.3 billion and delivered only a 20-ton payload to low Earth orbit. That's $65 million per ton—more expensive than gold or platinum.

Traveling to the stars requires vastly superior rocket technology. Antimatter fuel is the ultimate propulsion, a billion times more effective than chemical rockets. What would be possible if humanity could harness antimatter and how would Special Relativity affect us then?

Special Relativity limits a rocket's speed. As discussed above, objects appear to become heavier as their speed increases. Einstein's equations show that masses become infinite at the speed of light. Hence it would require an infinite amount of energy to accelerate a rocket, or anything else that has mass, up to the speed of light. As there isn't an infinite amount of energy in the entire observable universe, traveling at light speed or faster is impossible. This is well confirmed experimentally. In my own experience, the fastest massive particles ever produced by humans had velocities of 0.999,999,999,7c and masses 40,000 times more than identical stationary particles. (That was at the Stanford Linear Accelerator Center, which has been effectively shutdown.) While 40,000 is a bit short of infinity, the mass increase curve is exactly what Einstein predicted and there's no indication it won't continue rising into the wild blue yonder. Even if there were a deviation from Einstein's theory at higher velocities, the measurements we already have pose insurmountable limitations—no spaceship will ever attain the velocities already achieved in particle accelerators.

Where Special Relativity helps space travel is in extending the lives of astronauts. Einstein showed that time runs slower in moving frames: for example, time runs 10 times slower in a spaceship traveling at 0.995c. At that speed, the astronauts would age 10 times slower, and we on Earth would observe them living 10 times longer, perhaps 800 years. But unlike Methuselah, the astronauts would perceive that they had a normal lifespan of 80 years. They would have no more heartbeats, or joys, or sorrows than their Earth-bound peers—they would just seem to live in slow motion, while their spaceship blasted through space at 668 million mph. Thus, they would travel much farther within their lifespan.

A good flight plan would be to accelerate at 1g (32 feet per second per second) until reaching the halfway point, then decelerate at 1g, coming to a stop at the destination. This provides a livable environment during the long flight, with normal "gravity", while reducing flight time. For destinations farther than 400 light-years, the rocket should stop accelerating when it reaches a velocity of 0.999,99c, and coast until it's time to decelerate. The reasons to limit the spaceship's maximum velocity are cost and the ever-present cosmic microwave background radiation ("CMB"). CMB photons outnumber the particles of matter by a billion to one. We on Earth see these photons at an equivalent temperature of 2.7K, or −455 °F. But at 0.999,99c, a spaceship would face a torrential headwind of CMB photons at an apparent temperature of 1000 °F.

With few ways to dump excess heat in space, there will be a limit to the velocity that a spaceship and crew can survive. My crystal ball says that limit is not above 0.999,99c.

With this flight plan, the following table offers some round trip space cruises we might someday enjoy. We could visit a destination 20 light-years away and return. During this voyage, we would have aged only 12 years, but 43 years would have passed on Earth. If your less adventurous twin had remained on Earth, you would be 31 years younger than your twin upon arrival, and would stay 31 years younger from then on. Our maximum velocity would be 0.996c and 233,000 tons of fuel would be consumed per ton of payload. To a destination 400 light-years away, 803 years pass on Earth while we age 24 years; the maximum velocity is 0.999,99c and 26 billion tons of fuel are consumed per ton of payload. To a destination 2000 light-years away, 4003 years pass on Earth while we age 40 years; the maximum velocity and fuel consumption per ton of payload is the same as the prior example because our spaceship coasts for 1600 light-years each way, but we'd need to carry more payload to survive the longer voyage. Recall that we're assuming the most efficient fuel possible: half matter and half antimatter. In 2007, scientists estimated the cost of producing antielectrons at $2500 trillion per ton. Congress isn't likely to fund such trips any time soon.

## EINSTEIN IN TRAFFIC

Einstein said objects appear to shorten as they pass by us at speeds approaching the speed of light. Here's a fanciful example. Einstein is driving his new ***Ferrari Relativistica*** on the 405 freeway in Los Angeles when he sees a traffic jam ahead. Like a typical LA driver, Einstein will go to great lengths to avoid using his brakes. Between two trucks in the right lane, he spies a gap leading to an off-ramp and judges that the gap is exactly the same size as his car. Knowing that fast-moving objects shorten, and wanting to "fit in" in LA, he floors the accelerator and heads for the gap at 90% of the speed of light—go Al.

| Space Travel Planner (Round Trips) | | | | | | |
|---|---|---|---|---|---|---|
| Dist. 1-way lt-yrs. | Crew ages years | Earth time years | Max. speed vs. c | Max. *g* factor | Fuel to Payld. | See Note 3 |
| 1 | 3.8 | 4.5 | 0.744 | 1.5 | 46 | 6 |
| 2 | 5.3 | 7.0 | 0.867 | 2 | 200 | 13 |
| 4 | 7.1 | 11 | 0.943 | 3 | 1150 | 33 |
| 10 | 9.9 | 23 | 0.986 | 6 | 20 k | 142 |
| 20 | 12 | 43 | 0.996 | 11 | 230 k | 482 |
| 40 | 15 | 83 | 0.9989 | 21 | 3.1 M | 1760 |
| 100 | 19 | 203 | 0.9998 | 51 | 110 M | 10,400 |
| 400 | 24 | 803 | 0.99999 | 200 | 26 B | 160 k |
| 1000 | 30 | 2003 | 0.99999 | 200 | 26 B | 160 k |
| 1000 +X | 30 +X/100 | 2003 +2X | 0.99999 | 200 | 26 B | 160 k |

Notes:

1. For destinations up to 400 light-years, assume acceleration of 1g (32 feet/second/second) to midpoint, then deceleration of 1g, stopping at destination, and repeat on return. Beyond 400 light-years, coast after reaching 0.99999c until time to decelerate.

2. Assumes 100% fuel efficiency, a billion-fold improvement over current technology. Fuel usage is shown in tons of fuel per ton of payload, with k for thousand, M for million, and B for billion (1000 million).

3. For one-way trips to same destinations, flight durations are half of those listed, and fuel usage is given in right most column.

4. Relativistic *g* factor shown for maximum velocity; spaceship's clocks run *g* times slower than Earth clocks.

5. Planning longer trips is easy using the lowest row. For a round trip to a destination 2000 light-years away (X=1000), the crew will age 40 years (30+X/100) while 4003 years (2003+2X) pass on Earth. Maximum velocity and *g*, and fuel-to-payload ratio do not change since spaceship will coast throughout the added distance.

Can Special Relativity predict Al's fate? In their reference frame, the truck drivers see Al's car shrink to 43% of its normal length, so Al should make it through unscathed. But in Al's reference frame, his car remains its normal length, and it's the gap that shrinks—he's headed for disaster. Who's right? What really happens?

The answer is that whether Einstein passes or crashes actually has nothing to do with his speed. Let's assume Einstein is heading for the gap in a straight line, in a direction we'll call X. From his perspective, the gap between the trucks has a certain width perpendicular to X, which we'll call G. He also sees his car having a certain width perpendicular to X, which we'll call W. He'll pass if his car's width is less than the gap (W<G). Special Relativity states that lengths shorten along the direction of motion, in this case X, but not in perpendicular directions. Hence regardless of Al's speed, the values of G and W won't change. The truck drivers agree,

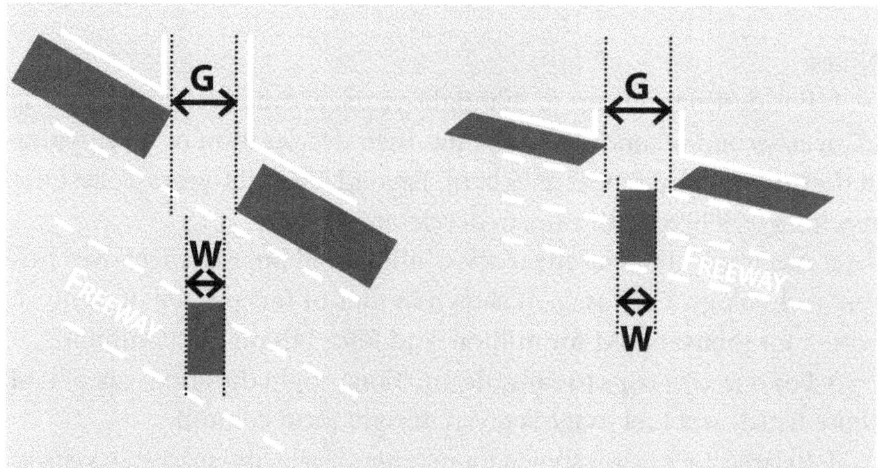


Figure 9.4. To avoid a traffic jam, Einstein steers his car toward a gap between two trucks and accelerates to 90% of the speed of light. Will he pass or crash? The left side of this figure compares G, the gap between the two trucks, with W, the width of Einstein's car, at low velocity. The right side shows how lengths parallel to the velocity contract at 0.9c, according to Special Relativity. Since they are perpendicular to the velocity, G and W do not contract. Einstein will pass if his car's W is less than G.

because they also see that G and W are in directions perpendicular to the velocity of Al's car. This is illustrated in Figure 9.4.

Of course, what happens to Al when he reaches the off-ramp at 90% of the speed of light is another matter.

This fantasy is a variation on a well-worn puzzle presented to students of Special Relativity: can a relativistic yardstick pass through a one-yard-wide window? The standard answer involves relativity's impact on the perception of simultaneity. That standard answer is bogus—it misses the essence of the problem, as Appendix A explains.

# 10

## General Relativity

Einstein's theory of General Relativity changed everything physicists believed about gravity and the universe, and superseded Newton's Law of Universal Gravitation. For the second time in a decade, Einstein revolutionized our understanding of space and time. General Relativity is Einstein's most profound achievement and his greatest contribution to science.

Special Relativity describes nature when everything is moving with a constant velocity. Since forces cause accelerations that change velocities, Special Relativity only strictly applies to ***special*** situations, where no forces exist or where their effects are small enough to be ignored. This limitation is not as restrictive as it might sound. Elementary particles experience forces only when they interact; between interactions Special Relativity applies. With some clever mathematics, Special Relativity is also adequate when accelerations are constant, such as the space travel examples in the last chapter.

But, ultimately we need a more comprehensive theory that can deal with forces, particularly as gravity acts everywhere throughout our universe. We need a theory that applies ***generally***, and this is what Einstein created in General Relativity.

## GRAVITY: NEWTON VS. EINSTEIN

Newton said gravity was a force that pulled together any two masses, such as the Sun and the Earth. He said every massive object produces gravity and responds to the gravity produced by other masses. He also said entities with no mass, such as light, neither produce nor respond to gravity. Newton's gravitational force was universal—the same law applied to all massive objects throughout the universe: stars, planets, moons, people, and even apples. For the first time in history, the human mind had mastered reality on a cosmic scale.

But Newton had no explanation for how gravity actually worked: how *did* the Sun reach out across 93 million miles of seemingly empty space and pull on Earth? That mystery bothered Newton, and all other physicists, but since his equations worked so well, everyone brushed aside the ***how*** and simply declared it was ***action-at-a-distance***, which sounded more professional than "ain't got the foggiest." Whatever action-at-a-distance was, it had to be infinitely fast—as a massive object moves, all other objects in the universe respond instantly to its new location, according to Newton.

Never one to sugar coat the truth, Einstein said every bit of Newton's theory was wrong. He said gravity wasn't even a force, but rather the effect of objects moving through a complex geometry: curved spacetime, which we will explore shortly. Einstein said it isn't just mass that produces the effect we call gravity, but rather all forms of energy, including mass-less electricity. And it isn't just mass that responds to gravity, but rather all forms of energy, including mass-less light. (Since Einstein had previously said mass and energy were equivalent, we should have seen this coming). Einstein also said action-at-a-distance doesn't exist—there are no invisible strings pulling things together. He said all action is ***local***—the Sun curves spacetime where ***it is***, and the Earth responds to curved spacetime where ***we are***; the fabric of spacetime is the mechanism that connects the two. Finally, Einstein said nothing, including gravity, travels faster than c, the speed of light; as energy and mass move, they change the local curvature of spacetime and these changes, called ***gravity waves***, ripple through the fabric of spacetime, moving outward at the speed of light.

Curved spacetime? The fabric of spacetime pushing Earth? Sounds like a psychedelic delusion—what was Al smoking? This was certainly a radical idea and a revolution in humanity's understanding of existence. But, having curved spaces control motion is actually something we've all seen before. In children's playgrounds, curved slides are big favorites (one is shown in Figure 10.1) as are giant roller coasters at amusement parks. Delighted

Figure 10.1. Motion is controlled by geometry. Here anything that enters the top of the slide must go round and round to reach the bottom because of the shape of this curved space.

riders are turned, spun, even hung upside-down. The shape of the space in which they are constrained to travel determines the riders' motions.

We are also familiar with other examples of moving through curved spaces. A ship just north of Antarctica will circumnavigate the globe if the captain steers due west. Even though the ship seems to move in a straight line, never changing its direction, it will travel in a circle and return to its starting point, as shown in Figure 10.2, because Earth's surface is a curved space.

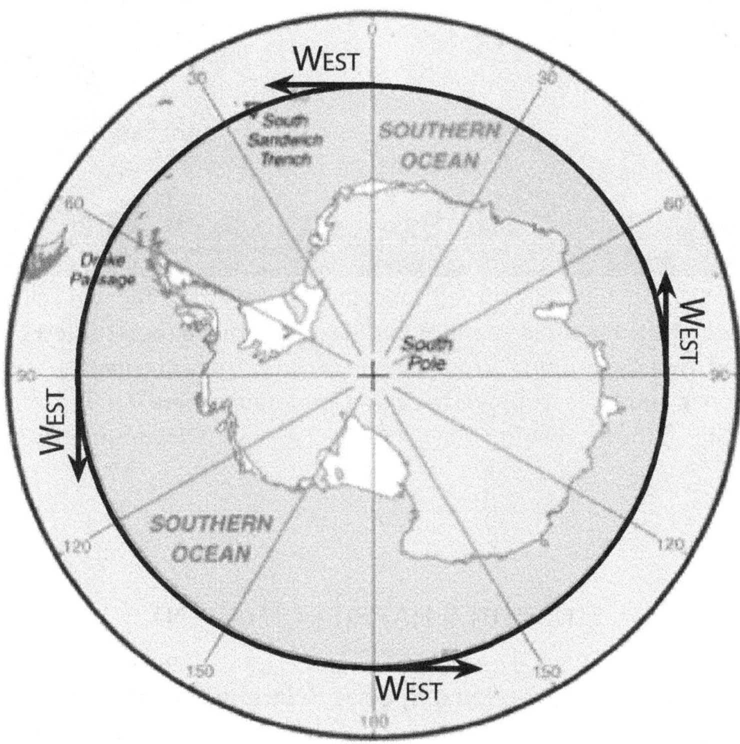


Figure 10.2. A ship heading due west will circumnavigate Antarctica and return to its starting point, because Earth's surface is a curved space.

An ant crawling along a corkscrew is similarly constrained by the shape of that space to go round and round in order to move from the shaft forward to the tip.

And in the same way, but in more dimensions than I can draw or even visualize, because of the geometry of spacetime in our Solar System, as the Earth moves forward in ***time***, it must go around the Sun. Figure 10.3 shows a representation of this effect in a reduced number of dimensions. The shape of space ***does*** control motion.

Figure 10.3. Einstein said curved spacetime causes the effect we call gravity. Here, the three dimensions of our Solar System are represented by the warped, two-dimensional, crosshatched surface. The Sun curves spacetime where ***it is***, and Earth responds to curved spacetime where ***it is***. The fabric of spacetime is the mechanism that connects them—not ***action-at-a-distance.***

## EINSTEIN'S HAPPIEST MOMENT

Soon after Einstein completed Special Relativity, he realized it would have to be extended to include gravity—he needed to develop General Relativity. But the problem seemed so difficult that almost no one thought it was possible; it seemed beyond the limits of the human brain.

In 1907, Einstein was still an obscure patent clerk. His great discoveries of 1905 had gone largely unnoticed—certainly they were unaccepted. With no prospect of a university position, he applied for a job as a high school physics teacher, but was summarily rejected. We now call 1905 his *Miracle Year*, but in 1907, it had yet to become a miracle for Einstein.

As the story goes, while staring out the window of the Bern patent office one day, Einstein noticed workmen on the roof of a tall, nearby building. The question occurred to him: if a person fell from a great height, what would they feel on the way down? (Let's ignore air resistance and avoid thinking about the unfortunate landing.) Einstein reasoned that a person in free-fall would feel ***no gravity*** whatsoever. He was correct, as everyone now knows from watching "weightless" astronauts on the International Space Station and in NASA's zero gravity simulator, affectionately dubbed the "vomit comet." But back in 1907, Einstein's insight was nothing short of brilliant, and he later called it his "Happiest Moment."

Einstein realized that since no forces act on an object in free-fall, that makes free-fall the most natural state of motion, where physics is simplest. The floor that pushes up against our feet and keeps us from falling makes physics more complex. "No forces" means we can use Special Relativity in the reference frame of a freely falling object. This made the esoteric mathematics of General Relativity much easier. With this simplification, Einstein took "only" another eight years to complete his theory.

Simple as it might sound, this is a profound idea with remarkable consequences, and thus deserves its elegant name: the Equivalence Principle. This principle states that the acceleration of gravity is equivalent, within a small local region, to a normal acceleration of the same strength. For example, imagine two sealed elevators. One elevator is stopped, with normal gravity pulling down with 1g on everything inside. The other elevator is in outer space, in zero gravity, with a rocket accelerating it upward at 1g. The Equivalence Principle says people would feel exactly the same effect in both elevators, having to resist equal forces pushing them toward the floor. And indeed, any scientific measurements made inside these elevators would yield exactly the same results. From inside, one cannot distinguish one elevator from the other. Gravity and acceleration are equivalent.

Einstein's theory of General Relativity encompasses gravity's three different effects: the normal acceleration of gravity, the tidal forces of gravity, and gravitational time dilation. These three effects of gravity have different characteristics and vary differently with distance, as explored in Appendix C.

## GRAVITY SLOWS TIME

Both Special and General Relativity predict that time flows at different rates in different situations. Special Relativity says time runs slower in moving systems, while General Relativity says time runs slower near sources of gravity. Experiments confirm both predictions.

How big is the second effect? Due to Earth's gravity, clocks at sea level run slower by 4 seconds per century than do clocks in some remote part of outer space where there is no gravity. Since all our clocks run slow by the same amount, no one notices. The Sun is 20,000 times farther from us than is Earth's center, but its mass is 300,000 times greater. The net effect is that the Sun slows our time even more, by 1 minute per century. Even the Andromeda galaxy "weighs" in on slowing our time. Andromeda is 15 million, million, million miles away but contains a trillion stars; it slows our time by 6 minutes per century.

Indeed, no part of our universe is remote enough to be totally isolated. Everything is interconnected through gravity.

## EINSTEIN MAKES GPS WORK

We said above that clocks on Earth's surface are slowed by 4 seconds per century due to Earth's gravity. We also said that no one notices, since all our clocks run slow by the same amount. That's true on Earth's surface, but not up in space. If the slowing of time due to gravity weren't properly accounted for, according to Einstein's General Relativity, GPS would be worthless.

GPS is the Global Positioning System operated by the 50$^{th}$ Space Wing of the U.S. Air Force. GPS provides precise navigation for cars, trucks, planes, and surface ships anywhere, anytime. A GPS receiver finds its position by measuring its distance from various GPS satellites orbiting 12,600 miles above Earth's surface. If the receiver can measure its distance from several satellites, its position can be determined by triangulation. GPS satellites continually transmit radio signals containing the current time and their orbital data. The distance to each satellite is found from the arrival time of its radio signal at the receiver, and the knowledge that radio waves are a form of light and hence travel at c, the speed of light. (Distance to satellite = c × reception time minus transmission time.)

The key thing here is getting the ***right time.*** A timing error of one millionth of a second results in a positioning error of 1000 feet. Because the satellites are 4 times farther from Earth's center than we are, their clocks are slowed 4 times less by Earth's gravity, only 1 second per century. Per General Relativity, their clocks run faster than ours by 3 seconds per century. At the same time, because the satellites are moving relative to Earth's center at 8640 mph, Special Relativity says their clocks are slowed by ¼ of a second per century.

The upshot of all this is that GPS would have a cumulative error of 6 miles per day without Einstein. So, if GPS got you home on Monday night, you'd be in the wrong city on Tuesday night, and out in the ocean by the end of the week.

If GPS gets you where you want to go, thank the man who could never find his own keys.

## PHYSICS: WHY GRAVITY SLOWS TIME

Let's understand why time runs slower closer to a massive body such as Earth. Imagine a laser at the top of the Leaning Tower of Pisa that fires a pulse of light toward the ground. Einstein said all forms of energy are affected equally by gravity, including light that has zero mass, because gravity is equivalent to acceleration. Anything that falls in a gravitational field gains energy—a rock falling down a mountainside gains kinetic energy as it moves faster and faster, while losing an equal amount of potential energy. The laser beam must also gain energy as it drops in Earth's gravitational field. Light cannot speed up, its speed is always c; instead it gains energy by increasing its frequency (more wave cycles per second). We'll adjust our laser so that the light pulse at the top of the Leaning Tower contains 400 trillion wave cycles and lasts 1 second, for a frequency of 400 THz (the reddest of red light). For this light to have a higher frequency when it reaches the ground, these 400 trillion wave cycles must occur in less time—if the frequency on the ground were 500 THz ("tera-Hertz"), the pulse must last only 0.8 seconds. If we measured the pulse duration at the top of the Leaning Tower with a clock at the top, and measured it again on the ground with a clock on the ground, we would measure 1 second at the top and 0.8 seconds on the ground. For the pulse to take less "time" on the ground, the clock on the ground must run slower. Time runs slower nearer Earth's center, where the gravitational potential is more negative. (To make the discussion simpler, I greatly exaggerated the frequency change; for the Leaning Tower, it would actually be trillions of times smaller.)

## BLACK HOLES

Black holes may be the strangest objects in nature, stranger than the most creative science fiction. They are the remnants of dead stars and the ultimate extreme of curved spacetime, only describable with Einstein's General Relativity.

Stars can exist for billions of years in an exquisite balance. In their cores burns nature's ultimate fire—nuclear fusion—producing unimaginable power. Our Sun, a fairly average star, produces 20 trillion times more energy than all of us on Earth consume. The Sun's energy creates an immense pressure that tends to blow it apart—340 billion times more pressure than we experience from Earth's atmosphere weighing down on us. Balancing the Sun's central pressure is its own gravity. Stars must be massive to avoid being blown apart; the least massive stars have 30,000 times more mass than Earth, and the most massive have 50 million times Earth's mass. Every bit of star pulls on every other bit with gravity. All these tiny tugs combine to produce an immense force that tends to crush the star. A star's gravity exactly balances its pressure. If your home thermostat were as well balanced, your house temperature would remain constant to 1/10,000$^{th}$ of a degree for billions of years.

Eventually, even stars die. Nuclear fuel runs out, fusion wanes, and the core pressure drops. Gravity begins to win the billion-year-long arm wrestling contest with pressure, and squeezes the core. As the core gets smaller, its self-gravity becomes rapidly stronger, and eventually, like an avalanche, the star's core collapses. Stellar collapses are among the most violent events in the cosmos, briefly increasing the star's power up to 100 billion fold. This vast energy release blasts away the star's outer layers and most of its mass, leaving only the collapsed core.

The cores of low-mass stars, the lightest 95% of all stars, will in death become ***white dwarfs***, about the size of Earth, with densities up to 30 tons per teaspoon. This will be our Sun's ultimate fate, some 5 or so billion years from now.

The collapsed cores of more massive stars will become ***neutron stars***, objects with more mass than our Sun, all compressed into a ball only 6 to 10 miles across, with densities up to 3 billion tons per teaspoon.

Finally, the collapsed cores of the most massive stars, the one in a thousand superheavyweights, become ***black holes***. Black hole masses range from a few times to 10 billion times the mass of our Sun—all of it compressed into a ball a trillion, trillion times smaller than a single atom. Historically, it was thought that this ball would be a single point with infinite density, called a ***singularity***. In physics, infinity is almost always a wrong answer. We now believe the ball is extremely small but larger than zero, and the density is extremely large but not infinite. However, the central object is generally still called a singularity.

To understand more about black holes, let's first discuss ***escape velocity***. If you toss a ball up into the air, it will slow down as it rises, eventually come to a stop, and then fall back down—that's gravity. But, if you could toss the ball up at 25,000 mph, it would slow down as it rose but would never come to a stop; it would be going fast enough to escape from Earth's gravity and fly off into outer space (and you'd have a fabulous baseball career). The speed required to escape a gravitational field, the ***escape velocity***, depends on the amount of mass creating the gravity, and the starting distance from the center of that mass. The escape velocity is greater if the mass is greater and is also greater closer to the center of that mass. Earth's escape velocity is greater at sea level than at the top of Mt. Everest, and is even greater at the Dead Sea.

So what does escape velocity have to do with black holes? While all its mass is in the central singularity, each black hole has another important feature: an ***event horizon***, where the escape velocity equals the speed of light. The event horizon isn't a physical object, but like Earth's horizon, it's the limit of how far we can see. Far away from a black hole, the escape velocity might be large due to its great mass—it might be 10% of the speed of light. Moving closer, the escape velocity increases. Where it reaches the speed of light is the location of the event horizon. This is the point of no return—a boundary that effectively separates the black hole from the rest of our universe. Anything that enters the event horizon can never get out because the escape velocity inside is more than the speed of light—nothing can move fast enough to escape, not even light. In that sense, the event horizon is an edge where part of our universe has been torn away, hence the word "hole." Since it neither emits nor reflects any light at all, black holes merit the other half of their name by being the blackest objects in the universe.

What would happen if someone fell into a black hole? As in any other free fall, the hapless victim will fall ever faster toward the center of gravity, in this case the singularity. The event horizon won't seem special in any way; he will pass right through it moving at 75% of the speed of light. Before reaching the singularity, the gravitational tidal force of the black hole will mercilessly stretch our victim from head to toe and crush him sideways—astrophysicists call this ***spagettification***. His body will be torn to bits, all of which will plunge into the singularity, where they will remain for eternity, we think. So our theories say, but since no one will ever be able to peer inside an event horizon and report back, we'll never be able to confirm what happens inside.

That is what we think our victim will experience in his reference frame. But his fate will look quite different to us on the outside. At first, we'll see the same thing—the victim falling ever faster toward the black hole. But as he approaches the event horizon, we will see him ***slow down*** and eventually come to a ***complete stop*** at the event horizon. How can that be? It's the slowing of time in strong gravity. From our perspective far away, time runs slower near the black hole and completely stops at the event horizon. Our victim's heart will stop beating and his brainwaves will stop waving. He won't exactly die, because death requires change and change requires the passage of time. But he won't exactly be alive either; he will be forever frozen in unchanging time. His image will rapidly become ever redder and fainter, and will soon be too red and too faint for us to detect. But in principle, it will never entirely disappear, remaining forever frozen on the event horizon along with the images of everything else that has "fallen into" the black hole.

## EINSTEIN'S FIELD EQUATIONS

General Relativity, the theory that gravity is the effect of curved spacetime, is expressed in Einstein's Field Equations:

$$G = 8\pi T$$

These equations (16 equations lie within these 5 symbols) are undoubtedly Einstein's greatest contribution to science. G represents the changing geometry of spacetime, and T represents the density of all forms of mass, energy, and pressure. As American physicist John Archibald Wheeler described, the Field Equations say: "Mass and energy tell spacetime how to curve, and spacetime tells mass and energy how to move."

While Newton viewed space and time as a fixed stage on which the drama of the universe is played out, Einstein envisioned a dynamic stage controlling the actors' motions and participating equally in the cosmic drama.

Brian Greene described the Field Equations as "The choreography of the cosmic dance of space, time, matter, and energy." Many scientists say General Relativity is the greatest scientific discovery ever made, while others go even farther and call it the greatest achievement of human thought. I believe General Relativity is a pinnacle achievement of human culture to be cherished along with our greatest achievements in music, literature, art, medicine, and other field of intellectual endeavor.

Einstein presented the Field Equations of General Relativity at a lecture in 1915 and in print in 1916. The theory gained wide acceptance when one of its remarkable predictions was confirmed in 1919. That year, during a total solar eclipse, English astronomer and physicist, Sir Arthur Eddington led two expeditions to opposite shores of the South Atlantic to test Einstein's prediction that starlight is bent by the gravity of the Sun. In Newton's theory, gravity does not affect light since it has no mass. Eddington's observations could only be made during a total solar eclipse since that's the only opportunity to view starlight that passes near the Sun. Total solar eclipses occur on Earth because of the peculiar coincidence that from Earth our moon has the same apparent size as the Sun—both subtend an angle of ½ degree—more on that later.

Eddington presented his results at a meeting of Britain's Royal Society. Standing under an immense portrait of Sir Isaac Newton, the patron saint of British science, Eddington proclaimed that Einstein's theory of gravity was right and Newton's was wrong. As the story goes, someone asked Eddington if it were true that only three people in the world understood General Relativity. He hesitated and eventually replied: "I was wondering who the third might be."

I noted above that total solar eclipses are visible on Earth due to a fortuitous coincidence—our moon's diameter divided by its distance from Earth just happens to nearly equal the Sun's diameter divided by its distance from Earth. Since the moon's orbit is elliptical rather than circular, its distance from Earth and its apparent size change. Thus the moon sometimes appears a bit larger than the Sun and sometimes a bit smaller. While our Solar System contains over 160 moons, none matches the Sun's apparent size as well as ours. We believe only on Earth is there a solid surface where total solar eclipses are visible—how lucky is that?

Imagine that intelligent life arose on another world and that someone there proposed the theory of General Relativity. Would scientists there accept General Relativity if they had no solar eclipses to confirm the bending of starlight? Maybe not. Testing General Relativity requires precisely measuring small deviations from Newtonian gravity. Scientists are properly skeptical of great claims whose supporting evidence comes from challenging experiments subject to many complex considerations. We wonder if something has been overlooked, or simply done wrong. Alexander Pope said: "To err is human; to forgive, divine." Scientists reviewing other scientists' work ardently believe the first half of Pope's dictum, but rarely practice the second half. When observational data must be extensively "massaged" to tease out a very small signal from a much larger background, there's lot of opportunity to err. Of all the tests that confirm General Relativity, the bending of starlight is by far the cleanest—take a photograph during a total solar eclipse and compare it to a photograph of the same area of the sky taken at another time. The bending angle is very small, but the procedure is straightforward and the required precision is well within the state of the art. Such photographs convinced Earth's scientists that Einstein was right. Without them, the debate would probably have raged on for many decades.

Indeed, of all the worlds in the universe that might support life, only a small portion might be able to support astronomers. For example, any life in oceans beneath the frozen surface of Jupiter's moon Europa would never see the heavens. Earth is remarkable, not just for supporting life, but also for supporting life that reaches for the limits of the universe.

## EINSTEIN AND COSMOLOGY

Einstein's General Relativity completely changed cosmology, our understanding of the universe. His new theory of gravity led directly to a new theory of an ever-expanding universe—the Big Bang theory—a concept that Einstein passionately rejected, but was ultimately forced to accept.

Shortly after publishing General Relativity, Einstein realized that when applied to the whole universe, his new theory did not allow a static universe. His equations required a universe that was constantly changing. This was contrary to a deep-seated belief of Western thought, which Einstein shared, that the universe is now as it always has been and forever shall be. With the benefit of hindsight, this belief is surprising. When we throw a ball up into the air, it rises, briefly stops, and then falls back down. Only for an instant, is the ball stationary and unchanging. Gravity demands change; everything in the cosmos moves.

Nonetheless, Einstein was convinced that the universe as a whole must be unchanging and he proceeded to fudge his equations by adding an arbitrary parameter, which he called the ***cosmological constant*** $\Lambda$ (the Greek capital letter Lambda). His concept, published in 1917, was that while gravity was pulling everything together, $\Lambda$, an explained property of spacetime, was pushing back with exactly enough strength to counter gravity and keep everything stationary. This wasn't Einstein's greatest moment—the cosmological constant made even less sense than action-at-a-distance or luminiferous ether. Being only a single number, the very best that Λ can do is balance gravity on an overall basis at a single instant in time. This is like balancing a knife on its point—it's an unstable equilibrium that can't persist—once the knife tips ever so slightly, it will promptly fall down. As planets orbit stars, stars orbit galaxies, and galaxies orbit clusters, masses move and gravity will pull in different directions and by different amounts. Inevitably, there will be more dense regions, where gravity is stronger than Λ. These regions will begin collapsing. As they do, their gravity will strengthen, making their collapse irreversible. At the same time, other regions will be less dense than average; there Λ will over-power gravity, causing expansion. As these regions expand, their gravity will weaken, making expansion irreversible.

In 1922, Russian mathematician Alexander Friedman published a solution of Einstein's Field Equations proving that a static solution was impossible—the universe either had to be expanding or collapsing; it could not be static. Starting in 1927, Georges Lemaitre, a Roman Catholic priest and physicist in Belgium, published papers and gave scientific talks expounding and gradually refining what would become the Big Bang theory. Early on, Einstein is reported to have told Lemaitre: "Your mathematics may be correct, but your physics is abominable." Einstein strongly believed in the power of physical intuition (particularly his own), and felt Lemaitre's idea was "ugly" and incompatible with nature's true beauty. In 1929, American astronomer Edwin Hubble discovered that all distant galaxies are moving away from us, and that the more remote galaxies are, the faster they are receding. The inescapable conclusion is that the universe is expanding. Einstein finally conceded, calling the cosmological constant "my greatest blunder."

It's OK to make some mistakes, provided one learns from such experiences. Explorers often make more mistakes than most, and Einstein made his share. But his errors didn't retard the advance of science; others promptly stepped in to set things straight. As Einstein once said: "Show me a person who has never made a mistake, and I'll show you someone who has never tried anything new."

Another major discovery paved the way for the acceptance and further development of Lemaitre's Big Bang theory of cosmology: the ***cosmic microwave background radiation***, or CMB. The CMB is a 13.7 billion-year-old relic—the afterglow of the primordial fireball—and strong evidence that our universe was once vastly smaller and hotter. Analysis of the CMB provides much vital information about our early universe. It shows that the universe was extremely homogeneous, having the same temperature everywhere to one part in 100,000. It also shows that, while spacetime is curved near massive bodies, the overall geometry of our universe is Euclidean—it obeys the rules of plane geometry that we all learned in high school.

We know the timeline of the evolution of our universe because of Einstein's General Relativity and the CMB. From the CMB, we know our universe is homogeneous and Euclidean. Remarkably, Einstein's Field Equations have ***only one solution*** for a universe with those two properties. Newton's theory has an infinite number of solutions and we might never know which one pertained to our universe. But since Einstein linked mass-energy with spacetime, only one solution is possible. This solution is named after its developers: Friedman, American Howard Percy Robertson, and Englishman Arthur Geoffrey Walker. Considering that it is the equation for the whole universe, the ***FRW equation*** is really quite simple:

$$3H^2 = 8\pi e$$

Here H is the Hubble expansion rate and e is the energy density of the universe, both of which change over time. With the CMB and other observations, we can determine the values of H and e today. Then using the FRW equation, we can compute the size and temperature of the universe at any time from an infinitesimal fraction of a second after the Big Bang to untold trillions of years into the future.

The cosmic microwave background radiation, the FRW equation, and Einstein's Field Equations are the cornerstones of modern cosmology and the Big Bang theory.

## THE BIG BANG THEORY

The Big Bang theory says our universe was born in an instant—call that instant ***t=0***—when its space, time, matter, and energy came into existence. Science is still trying to develop compelling answers to ***why*** and ***how*** this happened, and ***what preceded it***, if anything. The Big Bang theory describes what happened after ***t=0***. It says our universe was initially unimaginably hot and vastly smaller than the space now occupied by a single atom. Ever since, our universe has been expanding and cooling. At first, all matter was in the form of elementary particles that were too hot to stick together to build anything larger. But, as the universe cooled, particles coalesced to form small nuclei (at about ***t=100 seconds***) and to eventually form atoms, stars, galaxies, planets, and life.

When the universe cooled down to 3000 Kelvin (5000 °F), protons and electrons were cool enough to stick together and form hydrogen atoms. What had been a soup of charged particles that absorbed light, a ***plasma***, became a gas of neutral atoms through which light could freely pass. The primordial fireball became transparent, allowing light to propagate freely throughout the cosmos for the first time—this is the CMB that we see today. In our labs, we can measure the plasma-hydrogen transition temperature: it is 1100 times higher than the temperature of the CMB today, 2.7 Kelvin. As the universe expands, the wavelengths of CMB photons stretch and their equivalent temperatures drop. For the temperature to drop by a factor of 1100, the universe must expand by the same factor of 1100 in each of the three dimensions of space. From the FRW equation, we can calculate when the universe became transparent: 13.75 billion years ago, at ***t=380,000 years***. The age of the universe is now measured to a precision of better than ±1%—the uncertainty in its age is ±130 million years.

The first stars formed at about ***t=200 million years***. They were massive and therefore very short-lived. They produced the heavier nuclei, such as carbon, oxygen, and iron, which were dispersed into the cosmos when these stars died and exploded.

Our Solar System, the Sun and the planets, formed from a collapsing gas cloud enriched with these heavier atoms—the atoms that life requires—4.57 billion years ago, at ***t=9 billion years***. Radioactive dating has determined the age of our Solar System to a precision of 1%, meaning we know it is between 4.52 billion and 4.62 billion years old.

The earliest traces of life on Earth are 3.46 billion-year-old stromatalites—layered deposits possibly formed by blue-green algae. We believe multi-cellular life arose only 600 million years ago.

And here we are today, at ***t=13.75 billion years***.

For how long has this magnificent history of our universe been known? For less than 10 years, only since ***t=13,749,999,990 years***. How lucky are we to live in this Golden Age of cosmology?

# 11

# Quantum Mechanics

Quantum Mechanics is our theory of the micro-world of molecules, atoms, and everything smaller. Einstein made critical contributions to Quantum Mechanics, but never accepted its ultimate conclusions. Nonetheless, Quantum Mechanics has been outstandingly successful—its predictions have been tested by tens of thousands of experiments to extraordinary precision, and none of its predictions has ever been proven wrong. As strange as Quantum Mechanics may seem, one cannot deny that it describes the micro-world extremely well.

Quantum Mechanics is the foundation of all of chemistry, biology, and solid-state physics. The latter enables all the microelectronic devices that enrich our society. And yet, Quantum Mechanics is the most perplexing of scientific theories. Consider the comments of Niels Bohr and Richard Feynman, both recipients of the Nobel Prize for their major contributions to developing this theory. Bohr said: "If Quantum Mechanics hasn't profoundly shocked you, you haven't understood it yet." Feynman remarked: "I think that I can safely say that no one understands Quantum Mechanics." While scientists do completely understand the rules and equations of Quantum Mechanics, almost no one is comfortable with why nature behaves as it does.

Quantum Mechanics is based on two key principles: ***duality*** and ***quantization***. We discussed particle-wave duality in chapters 4 and 5. Quantization is best illustrated by comparing a ramp with a staircase; see Figure 11.1. On a ramp, any elevation is possible—by selecting the right spot on the ramp, one can be at any elevation from top to bottom. Staircases are different—only elevations corresponding to the steps are possible. One can be as high as the first step, or the second step, but never in between. We can say that elevation on a staircase is ***quantized***; it must be one of a limited number of choices. We can similarly say that U.S. money is quantized in units of one cent—the amount of money in any transaction must be a multiple of one cent.

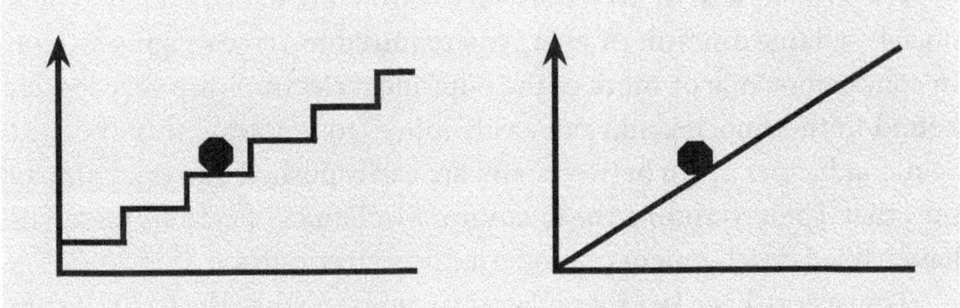

Figure 11.1. On a ramp any elevation is possible, as shown on the right. But on a staircase only a few specific elevations—the height of each step—are possible, as on the left. On a staircase, elevation is ***quantized.***

In the macro-world of our everyday experience, ramps are the norm. Planets, for example, can orbit their star at any distance at all. But in the micro-world of particles, quantization is common. Electrons can only orbit their nuclei at a limited number of very specific distances—their orbits are quantized. As discussed in chapter 5, an electron's orbit must be an integer multiple of its wavelength—the number of wavelengths is quantized in equal steps: 1, 2, 3… Each wavelength corresponds to a specific electron energy. Hence, electron energies in atoms are also quantized, but the energy steps are not of equal size.

## MICROELECTRONICS

Some materials, such as copper, readily conduct electricity. Other materials, like ceramics, are insulators that do not conduct electricity. Yet others are semiconductors that straddle this divide. Under the proper conditions, semiconductors can be rapidly switched at will between conducting and non-conducting states, thus providing outstanding control of electronic circuitry. Semiconductor switches, called ***transistors***, are trillions of times smaller, billions of times cheaper, and millions of times faster than mechanical switches, and have therefore revolutionized computing and countless related electronic applications.

All this technology is based on understanding the energies of electrons in various materials. In insulators, all electrons are tightly bound to their nuclei—a large amount of energy is required to remove any electron. In conductors, one or more of the outermost electrons are very loosely bound to their nuclei, and can easily move from one atom to the next. Semiconductors are in between, and are easily pushed into one camp or the other. Understanding the Quantum Mechanics of various materials has enabled development of modern microelectronics.

Transistors have largely replaced vacuum tubes as the basis of computing and electronic control systems. While vacuum tubes cost ten to several hundred dollars each, four billion transistors now cost just one dollar. Computing power that in my youth filled a building and cost tens of millions of dollars is now put to shame by microprocessors that are one-inch square and cost ten dollars. As a result, electronic computing has permeated nearly every aspect of our lives, generally for the better.

My toothbrush tells me how long to brush.

The computers that so many of us own are each composed of billions of transistors and other quantum mechanical devices. They have revolutionized our lives by allowing us to: shop for bargains online; communicate with friends via email and video; organize our finances; edit, process and share digital photos and videos; compose documents such as this book; read e-books; provide access to almost all human knowledge to everyone world wide via the Internet; and facilitate revolutions that liberate entire nations.

Digital microelectronics have also enabled cell phones, smart phones, and a variety of electronic "pads", which bring data, voice, video, and an entire e-world to our palms, putting to shame Dick Tracy's fantastic radio-watch.

Computerized robots perform delicate surgeries under a surgeon's direction, with precision that no human hand can match.

Electronics optimizes the operation of car engines, reducing emissions and fuel consumption.

Computed Tomography (CT), Magnetic Resonance Imaging (MRI), and other electronic medical devices scan our bodies for disease, without the need for surgery.

Microelectronics in cars can sense if you have adequate stopping distance, can navigate to your destination, and can automatically parallel park once you arrive.

Dictionaries and encyclopedias are nearly obsolete now that everyone has instant access to much larger online information "libraries."

The computerization of industrial production has dramatically strengthened our economy, even though it may "fly below the radar" of most people. Robots lift heavy parts and assemble them more precisely than humanly possible. Automated machines perform hazardous jobs, reducing injuries and deaths. Computer-aided-design (CAD) speeds product development and reduces errors. Computer-aided-machining (CAM) produces parts that are better and cheaper than any person could make.

Computer-guided laser systems can rapidly produce prototypes of new designs of any shape and configuration. This technology, based on two of Einstein's discoveries, accelerates new product development and reduces costs and mistakes.

## BARRIER PENETRATION

One of Quantum Mechanics' more bizarre predictions is that particles can pass through forbidden regions and emerge unscathed on the opposite side.

We know Earth's gravity pulls everything ***up here*** toward Earth's center ***down there***. Objects naturally settle down to their lowest possible energy level. They are driven to do so by the rules of probability that are enshrined in the Second Law of Thermodynamics: "Total disorder (what physicists call ***entropy***) shall increase." Thus, it's natural to see a rock roll down a mountainside to a valley below. But, what if an object lies at the

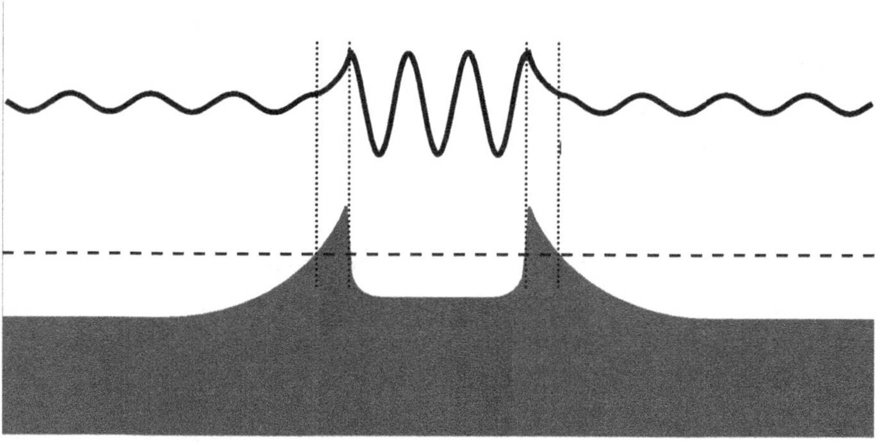

Figure 11.2. Quantum rules permit barrier penetration—particles can seem to pass through seemingly insurmountable barriers. Here the dark profile reminiscent of a volcanic caldera represents a particle's potential energy near a radioactive nucleus at the center. The dashed horizontal line represents the total energy (kinetic plus potential) of a particle originally inside the nucleus. The total energy is insufficient to scale the barrier. But, the particle's wave, shown at the top, doesn't stop abruptly at the barrier; it smoothly diminishes as it penetrates the barrier, waving again once outside. Because the wave amplitude at any point is related to the probability of finding the particle at that point, the particle may suddenly appear outside the nucleus.

bottom of a volcanic crater, a caldera, surrounded by steep walls? The valley beyond may be lower than the caldera floor, as in Figure 11.2,but those steep walls will trap the object inside. Right? Yes, if the object is a rock. No, if it's an electron.

Recall that Einstein discovered that light has both particle and wave properties, and that de Broglie extended this idea, establishing the concept of particle-wave duality—everything is both a particle and a wave. Thus particles have wavelengths, which are generally negligible in the macro-world but are critically important in the micro-world. In a micro-world caldera, an electron wave will be primarily within the steep walls. But its wave doesn't stop dead at a wall—waves are smooth and can't drop instantly to zero amplitude. Quantum Mechanics says the wave's amplitude (the height of its crests and depth of its troughs) will decrease rapidly but smoothly as it penetrates the wall. The higher the wall, the more rapidly the amplitude drops, and the thicker the wall, the smaller its amplitude gets—it can become extremely small, but it never drops to zero. Outside the caldera, the wave resumes waving, but with the much-reduced amplitude; see Figure 11.2. An electron wave's amplitude at any point determines the probability of finding the electron at that point. This means there's a small chance that the electron will suddenly appear outside the caldera, as if it had penetrated the wall, a seemingly insurmountable barrier. Quantum rules do not permit the electron to appear within the barrier—it will almost always remain inside, but if the walls are not too high and not too wide, it may suddenly appear outside.

Almost every electronic device on the market relies on the ability of electrons to penetrate barriers in defiance of pre-quantum rules.

Radioactivity also relies on barrier penetration. In very large nuclei, each proton is repulsed by every other proton and attracted by only a few, because the electric force has an unlimited range and the strong force has a very short range. A nucleus is unstable if some of its particles have too much energy. In the volcano analogy, this occurs if the particles' total energy is higher than the valley beyond the caldera. Eventually, part of the nucleus will appear outside—the nucleus will *decay*.

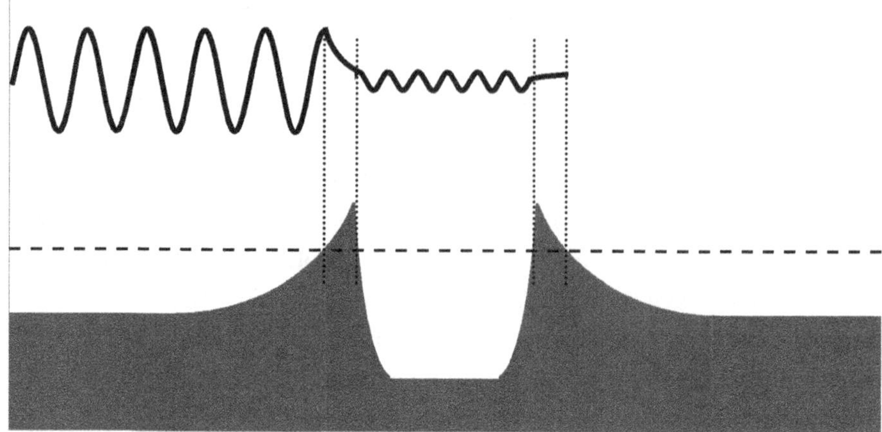

Figure 11.3. Nuclear fusion is enhanced by barrier penetration. Here the caldera floor is lower than the region outside, indicating the lower energy that two nuclei will have after merging. The dashed horizontal line represents the total energy of a nucleus (kinetic plus potential) approaching from the left. Barrier penetration enables a small portion of the incoming nucleus's wave to enter the nucleus at the center. The two nculei can then merge even if the total energy is insufficient to overcome the repulsion of their positive electric charges.

Nuclear fusion also relies on barrier penetration, but in reverse; see Figure 11.3. Returning to the volcano analogy, imagine now that the caldera floor is much lower than the valley beyond the volcano. This is similar to the situation of two nuclei merging in the process of nuclear fusion. The electric force strives to keep positively charged nuclei apart—that's represented by the steep walls surrounding the caldera. But if the nuclei are able to come together, the strong nuclear force will bind them extremely tightly—that's represented by the deep caldera floor, far lower than the top of the wall and even lower than the valley beyond. Because nuclei have wavelengths, their waves can penetrate walls, allowing nuclei to suddenly appear inside the caldera. This enables nuclei to fuse even if their energy is insufficient to completely scale the walls of electric repulsion. If not for barrier penetration, stars would be vastly dimmer and our Sun could not sustain life on Earth.

## WHAT IS SUPERCONDUCTIVITY?

In chapter 7, we explored the antisocial behavior of fermions, the particles of matter that include electrons. This is manifest in each electron refusing to share and demanding its own turf. (Reminds me of my two-year old grandchild who thinks sharing is highly overrated.) Recall this is due to electrons having a quantum spin of ½. However, Bose and Einstein showed that bosons, objects with integral spins (0, 1, …), are gregarious and are eager to share their turf. This is particularly important at low temperatures, where such particles are cool enough to stick together and form ***Bose-Einstein condensates.*** American physicist Leon Cooper took this a step further with the clever idea that, at low temperatures and in the right materials, two electrons can pair up, making a combined object with spin 0 or 1, thus becoming a boson. Normally, electrons in conductors are antisocial and impede each other's movements, which causes resistance to the flow of current. Resistance wastes power and creates unwanted heat, sometimes causing fires. But ***Cooper pairs*** are gregarious bosons that share their quantum state with similar pairs, allowing them to flow freely through conductors with absolutely no resistance at all—we call this ***superconductivity.*** Cooper, John Bardeen, and John Schrieffer received the Nobel Prize for discovering how this works.

## SUPER USES OF SUPERCONDUCTIVITY

While superconductors remain rather expensive, they offer unmatched performance and are sometimes the only practical solution in demanding applications. Today, most applications of superconductivity involve high-strength magnets that are possible only using superconductors. By eliminating electrical resistance, superconductors can carry much greater electric currents without power losses and damaging heat, creating extremely powerful electromagnets. These magnets pack more power in less space than any other technology.

Superconducting magnets are used to levitate trains above their tracks. These trains effectively fly just above the railway with absolutely no contact, thereby eliminating rolling friction and delivering a smoother ride. On the Yamanashi Maglev Line in Japan, a test train has achieved 361 mph.

Another application of high-power superconducting magnets is MRI, an advanced non-invasive medical imaging modality. MRI scans show doctors how various parts of our bodies are functioning, even parts that are deep inside us. MRI accomplishes this without ionizing radiation and without surgical incisions.

The most powerful particle accelerators for fundamental physics research also rely on superconducting magnets to keep their particle beams going around on circular paths. These accelerators are the world's most powerful microscopes, exploring nature at distances of one billionth of one billionth of an inch.

SQUIDs, Superconducting Quantum Interferences Devices, can detect magnetic fields a billion times weaker than Earth's field. SQUIDs are used by the U.S. Navy to detect the stray fields in submerged metal objects such as mines and submarines.

Electric motors and generators made with superconductors provide more power, more efficiently, in less space than conventional devices, and are just now entering industrial use.

Our existing non-"super" power grid can only transport electricity about 500 miles from power plants to consumers. This limitation is due to excessive radiative power losses from 60Hz alternating current over longer distances. This isn't a severe limitation for fossil-fuel power plants that can be built almost anywhere. But to expand solar and wind energy production, we need a transmission system that spans the nation. We need to be able to efficiently move power from solar plants in Arizona and wind turbines in the Dakotas to consumers thousands of miles away. The best solution may be superconductors carrying direct current. Direct current does not radiate energy as alternating current does, and with superconductors, DC power can be transmitted without excessive resistance, power loss, and heat.

# 12

# Einstein as a Leader of Science

Great quarterbacks do more than just throw amazing passes—they lead. They set the right example, unite diverse talents and interests, identify common and compelling goals, and focus a team's efforts on achieving success. Einstein did all this for science—leading by example.

## ICON OF SCIENCE

We need role models and we need memorable images.

Time magazine chose Albert Einstein as their "Person of the Century", proclaiming him the "greatest mind and paramount icon of our age." Einstein will forever be everyone's image of the brilliant, but absent-minded, scientist who could fathom the universe but couldn't remember where he left his paycheck. Who can forget the explosion of white hair that never saw a comb, or his great equation $E=mc^2$, which adorns coffee cups and T-shirts across the world?

Einstein should also be remembered for his heroic spirit. He wasn't born to success; indeed, for the first 30 years of his life it seemed he was destined to failure and obscurity. Many others would have given up in frustration, but despite frequent rejection, Einstein never stopped trying. His first college application was rejected. His first Ph.D. thesis was rejected. After graduating, he was rejected at every university and technical institute in Europe. He was deemed unqualified to be a government clerk and to teach high school physics. Yet Einstein kept trying. He worked on several problems nearly non-stop for ten years or more—truly remarkable. It was his dogged determination, coupled with immense talent and considerable luck that ultimately turned abject failure into unparalleled success. Perseverance made Einstein the most famous scientist in history—the Elvis Presley of Physics. It is impossible to overstate the unparalleled esteem accorded Einstein by scientists, the public, and the leaders of society.

## UNIFICATION: THE PRIMARY GOAL

Einstein had an unmatched ability to perceive the fundamental unity between things that seem totally different. In doing so, he reduced the complexity of science and brought new and more profound understandings of nature. Einstein believed nature is fundamentally simple, harmonious and beautiful. When it looked complex, he believed it was because we had not yet seen its underlying simplicity.

Einstein's search for nature's unity and simplicity dominated nearly every waking moment of his life. Through his leadership, Einstein made this search the holy grail of physics. Even half a century after his death, thousands of physicists around the world jump out of bed every morning, eager to pursue the course that Einstein set.

In 1905, Einstein published one of the most important papers in the history of science, his Theory of Special Relativity. The paper begins by highlighting something Einstein considered ugly and unacceptable: Electromagnetism had two equations for one problem. According to a "constant field" equation, if a wire moves past a stationary magnet, the unchanging magnetic field causes an electric current to flow in the wire. But accordingly to a "changing field" equation, if a magnet moves past a stationary wire, it creates an electric field that causes a current to flow in the wire. Both equations yield the same current, so no one much cared—no one except Einstein. To him this was ugly and he knew nature wasn't

Figure 12.1. Time Magazine chose Albert Einstein as their "Person of the Century", saying he "...had the greatest mind and was the paramount icon of our age."

ugly. The Principal of Relativity says it is impossible to know whether the wire moves past the magnet or the magnet moves past the wire; only relative motion is physically meaningful. Hence, Einstein said having two equations for one problem means we don't fully understand nature—one phenomenon must be explained by one theory (not two) and be solved by one equation (not two). Mathematical "ugliness" seems to have been more important to Einstein than any laboratory measurement, and it inspired him to search for and discover a more profound understanding of Electromagnetism and motion: Special Relativity.

Einstein discovered that particles and waves were not completely different phenomena as everyone else believed at the time. He showed that waves sometimes act like particles, so they cannot be entirely different. Eventually his discovery led to the concept of particle-wave duality—that "particle" and "wave" are merely labels for two ends of a spectrum, like "black" and "white." Everything in our universe is a shade of gray; everything has a mixture of particle and wave characteristics.

Einstein discovered that mass was a form of energy even though it appears to be completely different from all other forms of energy, such as heat. He further declared in his most famous equation $E=mc^2$ that mass was not conserved, as everyone else believed at the time, but that it could be converted into other forms of energy.

Einstein showed that space and time were not separate and absolute entities, as Newton believed, but were united as a single entity: spacetime. He showed that different observers, moving at different speeds, would necessarily make different measurements of space and time—space and time are *relative*. One person's space might be another person's time.

Einstein showed that the inertial mass in Newton's motion equation doesn't just happen to be numerically equal to the gravitational mass in Newton's gravity equation. He showed that these two masses must be identically the same because there is only one type of mass. Einstein said there is no "force of gravity"—things move as they do because spacetime is curved. All objects, he said, move along the same paths through curved spacetime regardless of their masses. There is no need for two mass types.

Finally, Einstein showed that spacetime and mass-energy are also intimately interrelated. Spacetime is the manifestation of "gravity" due to the presence of energy, including mass. Spacetime and mass-energy cannot exist without one another—they are two sides of one coin. Spacetime tells mass and energy how to move and mass and energy tell spacetime how to curve—it is a cosmic duet.

A more profound understanding of science is important for two reasons. First, advancing science expedites discoveries that enrich our society technologically and financially. Second, science is culturally important, like art and literature. Romeo could have said: "Names are like whatever. Roses smell good, totally." We prize Shakespeare because he expressed profound truths beautifully and memorably. William and Albert both advanced our culture, set higher standards, and inspired countless others to do their very best to emulate their achievements.

# 13

## If Not Einstein, Who and When?

At the start of this book, we posed an intriguing question: If Einstein had not existed, wouldn't someone else eventually have made the same discoveries?

Of course, we can never know for sure what would have happened if history had been different. Most advances in science and technology are modest, incremental improvements, even those rewarded with patents and prestigious prizes. Many people have similar talents and could have made these same advancements, given enough time and support. Were Einstein's contributions merely slightly better mousetraps? Could others have filled Einstein's "shoes"?

To a substantial degree, I think not. While some of his lesser works were within the scope of other researchers, many of his contributions were truly extraordinary.

If Columbus had not discovered America, someone else would eventually have crossed the Atlantic, because ship technology was steadily advancing, driven by economic motives that had nothing to do with discovering America. But Einstein's discoveries weren't driven primarily by a general advance of science and technology. His discoveries came from his own thoughts and his resolve to understand fundamental questions that scientists had struggled with since the dawn of recorded history: light, matter, energy, space, and time.

Einstein was surely the greatest scientist of the last three centuries—no serious person could disagree. Experts might debate whether the contributions of Newton or Einstein are the more profound. But certainly, these two physicists stand by themselves on the highest rung of science—they are unrivalled going back at least as far as Aristotle over 23 centuries ago. While a sample of two isn't statistically robust, the best one can estimate is that humanity produces an Einstein less often than once every 1000 years. That alone speaks volumes.

But genius alone does not guarantee recognition and achievement. As remarkable as Einstein's early papers were, the real Miracle of 1905 was perhaps that any major science journal published them. Einstein offered no proof for his revolutionary theories—verification came from experiments performed decades later. Additionally, Einstein's analytical methods were so unconventional that many physicists had trouble understanding what he wrote, let alone deciding to believe any of it. And who was this Einstein fellow anyway? He was not a university professor, did not have a Ph.D., and did not work at an established science institution. It's doubtful that such radical papers from an uncredentialed author would survive today's bureaucratic peer-review process. In 1905, the decision to publish Einstein's papers was made by a single individual, Max Planck, editor of physics' most prestigious journal. Planck didn't believe much of what Einstein wrote, but nonetheless found his ideas interesting. Later, it was Planck who convinced the world's leading university to hire Einstein, and it was Planck who recommended him for a Nobel Prize. Had these papers fallen into less enlightened hands, Einstein's great discoveries might never have been noticed. We will never know how many other brilliant ideas have been ignored.

Now let's consider the specifics of Einstein's achievements.

If President Roosevelt had not acted on Einstein's recommendation, World War II would not have ended as quickly as it did. No scientist in the world (then, before, or since) enjoyed the unquestioned prestige of Albert Einstein. Even if another scientist had made this recommendation, and none did, it is highly doubtful that the U.S. would have invested so much so quickly. Einstein's theory, his decision to act, and his unequalled stature were essential in saving millions of lives.

In 1905, Einstein devised a method to prove that atoms really do exist. The world's leading scientists had vigorously but unsuccessfully debated their existence for 2500 years. Einstein thereby solved the mystery of Brownian motion that had baffled scientists since observations by Brown in 1827 and Ingenhousz in 1785. Those observations predate Einstein's solution by 78 and 120 years, respectively—for all those years, no one else had been able to explain Brownian motion. And for 2500 years, no one else had been able to prove that atoms exist.

Hertz observed the photoelectric effect in 1887, and Becquerel observed the related photovoltaic effect in 1839. No one succeeded in explaining these effects for 18 and 66 years, respectively. In 1905, Einstein published the solution to both mysteries: light is both a particle and a wave. Einstein's solution contradicted what all other scientists had firmly believed for hundreds of years.

In 1678, Dutch scientist Christian Huygens published his theory that light was a wave moving through an undetected substance—***luminiferous ether***. For over two centuries, physicists tried in vain to find this illusive ether, with the most decisive negative result by American physicists Michelson and Morley in 1887. Ether became one of physics' most important issues in the mid 19$^{th}$ century. Many eminent scientists devoted themselves to resolving the seeming incompatibility of two major theories: Electromagnetism and Relativity. Yet, in all that time, no one suggested Newton's Laws of Motion were wrong or that time and space were not absolute and universal. Instead, theories were concocted about how objects are compressed as they pass through ether. Some derived the right equations, but for all the wrong reasons. Einstein proved they were all pursuing blind alleys. If no one else was able to discover Special Relativity in one to two centuries, how much longer might this have taken without Einstein?

Another revolutionary departure from long-established science was Einstein's discovery, in 1905, that mass was a form of energy and that mass could be converted to or from other forms of energy. It is the total amount of energy in all forms, including mass, Einstein said, that is conserved. Einstein thus repudiated the principle that mass, by itself, is conserved. This principle was first enunciated by the ancient Greeks (Empedocles in the 5$^{th}$ century BC) and established as a major principle of modern science by French chemist Antoine Lavoisier in 1789. Lavoiser's scientific career was regrettably cut short by the guillotine during France's Reign of Terror—condemned by a "judge" who reportedly said: "The Republic needs neither scientists nor chemists." The conservation of mass was more than just a belief—it was well confirmed by all experiments of the day. Only in the 1930's, spurred on by Einstein's theories, did scientists develop the instruments and knowledge to successfully observe mass being converted into energy. If not for Einstein, his brilliant insight might have remained undiscovered for decades, as it had for centuries before.

Lastly, consider Einstein's greatest achievement: General Relativity. Newton's Law of Universal Gravity was the foundation of modern science—for the first time in history, the human mind could describe the behavior of everything. For over two centuries, Newton's equations were confirmed to unprecedented precision by countless observations. No one was looking for or expected a radically new theory of gravity. Similarly, Euclidean geometry had been a cornerstone of mathematics and science for 25 centuries. No one expected that to change or could have guessed that space and time are "curved"—a notion so complex that almost no one understood it, even after Einstein explained it. Some of today's leading physicists believe that without Einstein, we would still not have developed a theory of gravity as curved spacetime, nearly 100 years later.

In his Theory of General Relativity, Einstein predicted ***gravity waves***, ripples in the fabric of spacetime, caused by the motion of massive bodies. Sixty years later, American astronomers Russell Hulse and Joseph Taylor discovered two neutron stars that are very close and rapidly orbiting one another—their "year" is less than 8 hours long. Gravity waves carry away energy, forcing the neutron stars into ever-smaller orbits. Hulse and Taylor observed the orbital period decreasing at the rate of 76.6±0.8 seconds per million years, in complete agreement with General Relativity's prediction of 75.8 seconds per million years. Einstein predicted, with astounding precision, a phenomenon occurring 120,000 trillion miles away, something that no one had ever dreamt of and that would not be discovered for 60 years.

Could anyone else really fill Einstein's shoes? Perhaps one day, but so far no one has come close.

Einstein solved mysteries and overthrew cherished but incorrect principles that had reigned for decades, centuries, and even millennia. In all those years, someone else could have made these discoveries, but no one did. Consider again humanity's history of producing a scientist of Einstein's caliber less often than once every 1000 years. While we will never know for sure, it seems highly probable that without Einstein, the advance of science would have been severely retarded. And without Einstein, we would not yet have many of the modern technologies that we enjoy and that increase the wealth and wellbeing of our society.

One person truly can make an enormous difference.

# A

## Yardsticks and Windows

The fanciful example of "Einstein in Traffic" in chapter 9 is a variation on a well-worn puzzle presented to students of Special Relativity. Here is a less exciting but more common version: if a relativistic yardstick passes a window that is one yard wide, will the yardstick fit through the window? As the puzzle is presented, the yardstick's velocity is parallel to the window aperture. Special Relativity says that in the yardsticks' reference frame the window will shrink and block the longer yardstick, while in the window's reference frame the yardstick shrinks and passes through the wider window.

The standard answer to this apparent paradox is: the yardstick passes through the window, because relativity alters the perception of simultaneity, allowing one end of the yardstick to pass through before the other. That standard answer is completely inappropriate. Certainly it is true that events that appear simultaneous in one reference frame will not appear simultaneous in frames in relative motion. But this has nothing whatsoever to do with this puzzle.

If the yardstick's velocity were truly parallel to the window aperture, it would never pass through at any speed, relativistic or not, because it's going in the wrong direction. The key point here is that the yardstick's direction of motion must intersect the window aperture for the yardstick to have any chance of passing through it. The relativistic contraction will occur along that direction of motion, and not perpendicular to it. The yardstick can pass if it is narrower than the aperture in the two directions perpendicular to the motion. The puzzle's answer depends only on perpendicular distances, which are unchanged at any velocity. If properly aimed, a one-inch wide, six-foot long javelin can easily pass through a one-yard wide window—the javelin's length and velocity are irrelevant.

The true answer to this puzzle is that it has nothing to do with Einstein's theories of relativity.

# B

## Is Magnetism a Special Effect?

In chapter 9, we said that the effects of Special Relativity are extremely modest at normal velocities, those much below the speed of light. There is a remarkable exception that was pointed out to me by my brother, Dr. Richard Piccioni, who teaches high school physics.

One of many "gee whiz" experiments that first-year physics students enjoy is running electric currents through two parallel wires. When the currents are switched on, the wires will jump together if the currents flow in the same direction in both wires, and will jump apart if the currents are in opposite directions. Why? Normally, students are told this is due to ***magnetism***. That isn't wrong, but a deeper answer is: ***Special Relativity***.

At the atomic scale, an electric current is composed of a vast number of electrons "hopping" from atom to atom in a consistent direction. Any single electron "hops" extremely rapidly but very infrequently. The "average velocity" of the current-conducting electrons is a macro-world concept that doesn't fully describe the reality of the micro-world. Nonetheless, this average velocity, v, does usefully describe the current flow observed in the macro-world. The value of v is exceedingly small—typically about 1 foot per hour, or about 1/3 of a trillionth of the speed of light. At that velocity:

$$g = 1.000,000,000,000,000,000,000,000,05$$

How could such an itsy bitsy effect move large objects like wires?

Here is Richard's analysis. (See Volume 45, the March 2007 issue of "The Physics Teacher" for all the details.)

Let's start with two parallel wires, called wire A and wire B, which carry equal currents in the same direction. In an electrical conductor like copper, the outermost electron of each atom is loosely bound to the nucleus and can easily flow from one atom to the next. If an atom loses an electron, it will have a net positive charge—we call that type of atom an *ion*. In the laboratory reference frame, where we are stationary, we observe electrons moving with velocity v through both stationary wires with stationary ions. We also note that, like nearly every macroscopic object, each wire has zero net electric charge—the number of electrons per inch of wire equals the number of ions per inch. So far, there's no apparent reason for a force between the wires—but we're not done yet.

Now, consider how this appears in the reference frame of a current-carrying electron in wire A. Since velocity is relative, this electron "sees" itself and every other conduction electron as stationary, but observes both wires and their ions moving with velocity –v (same speed as above but in the opposite direction). Special Relativity says lengths shorten along the direction of motion. Hence the electron will see less space between the moving ions, yielding a higher number of ions per inch than we observe in the lab frame. The electron will also see a lower number of electrons per inch than we observe in the lab frame, which is a bit trickier. To understand why, recall that in the lab, the electrons are moving and thus their spacing appears smaller to us than it appears in the electron's frame. Turning that around, the electron spacing appears larger in the electron's frame than we observe in the lab. Thus, the electron sees more ions and fewer electrons per inch—the wires that appear uncharged in the lab appear positively charged to the conduction electrons. The excess positive charge in wire A doesn't affect the electrons in wire A since the same excess charge exists in every direction and those forces cancel. But the excess positive charge in wire B attracts the electrons in wire A, and *vice versa*, forcing the wires together.

Doesn't the excess positive charge in each wire force them apart? No, because this excess is only observed in the reference frame of the electrons. In the reference frame of the ions (the lab reference frame), the net charge of each wire is zero. Thus, the wires pull one another together. Even though the amount of relativistic shortening is incredibly small, there are so many electrons and ions per inch that the total force is substantial—enough to snap the wires together. This should give everyone of us a heightened appreciation for the immense number of atoms in every macroscopic object, even an inch of wire.

If currents in the two wires flow in opposite directions, the electrons in wire A still "see" an increased number of ions per inch in wire B. But now they also see the electrons in wire B moving at –2v, faster than the ions moving at –v. The observed increase in electrons per inch exceeds the increase in ions per inch, making wire B's apparent net charge negative, pushing the wires apart.

What is normally called a magnetic force is at a fundamental level the result of length contraction due to Special Relativity, even at an exceedingly low velocity.

# C

## Gravity and the 3 R's

Gravity seems simple—stuff falls down. But gravity actually causes three different effects that vary quite differently with distance. These effects are the normal acceleration of gravity, the tidal forces of gravity, and time dilation due to gravity.

The normal acceleration of gravity, a, is proportional to M, the mass of the gravitating body, such as Earth, and inversely proportional to the square of the distance, r, to the center of that body (if it's spherical). In appropriate units, the equation is: $a = M/r^2$. This is often referred to as an ***inverse-square*** law. Newton proved a marvelous theorem: for two spherically symmetric bodies, the acceleration of gravity doesn't depend on how either mass is distributed. We may assume both bodies have all their mass concentrated at a single point, which makes the calculation much simpler since every particle in one body is then at the same distance from every particle in the other body—there's only one value of r. So, all we have to do is add up all the particle masses and divide by $r^2$. If this weren't true, the process would be much more tedious: compute each particle-to-particle force with a different r and then sum that myriad of forces, all with different strengths and all pointing in different directions. Newton said an object's size doesn't matter and neither does its density; only its total mass counts. It turns out that the same thing is true in General Relativity: we may assume all of a symmetric body's mass is acting at its center.

Everything in the preceding paragraph is true *outside* a symmetric object. An interesting thing happens inside. At the bottom of South Africa's Tau Tona gold mine, 2.4 miles below Earth's surface, the acceleration of gravity is: $a = M^*/r^2$, where $M^*$ is the portion of Earth's mass that is closer to the center than r. Earth's mean radius is 3959 miles, thus $M^*$ is the mass contained within 3956.6 miles of Earth's center. What about the outer 2.4 miles? The acceleration due to those 3 billion, billion tons is zero!

Why? Imagine being inside a thin, spherical shell of material (like being inside a balloon). Let's calculate the acceleration of gravity at any arbitrary point inside, in any arbitrary direction, as illustrated in Figure C.1. Place the tip of a cone at the selected point and extend the base of the cone to the shell in the selected direction. Pick a cone with a very small opening angle, α. Now add a second identical cone going in exactly the opposite direction, with its tip touching the tip of the first cone and its base extending to the opposite side of the shell. Call the distance from tip to base in the first cone $r_1$, and the corresponding distance in the second cone $r_2$. In the first cone, its base area is $kr_1^2$ (with $k = \pi \sin^2 \alpha/2$). The amount of mass within the shell covered by the base of the first cone is then $dkr_1^2$, where d is the mass density per unit area. The acceleration due to that mass equals the amount of mass divided by the distance squared: $a = dkr_1^2/r_1^2 = dk$. Note that the distance cancelled—the amount of mass covered by the base of the cone increases with the distance squared and gravity's strength decreases with the distance squared, so the distance $r_1$ doesn't matter. The acceleration due to the mass covered by the base of the second cone is similarly: $a = dkr_2^2/r_2^2 = dk$. Since these masses lie in exactly opposite directions and are pulling with equal strength, they exactly cancel one another—no net force. And, since the direction of the cones and the starting point were arbitrary, this result applies in all directions from all points inside the shell. Thus inside any thin, symmetric shell, the acceleration of gravity is zero everywhere. A thick, symmetric shell is just the sum of many thin concentric shells, so the theorem applies to those as well, even if the thin shells have different mass densities. G. D. Birkhoff proved these special properties for symmetric masses are also valid in General Relativity.

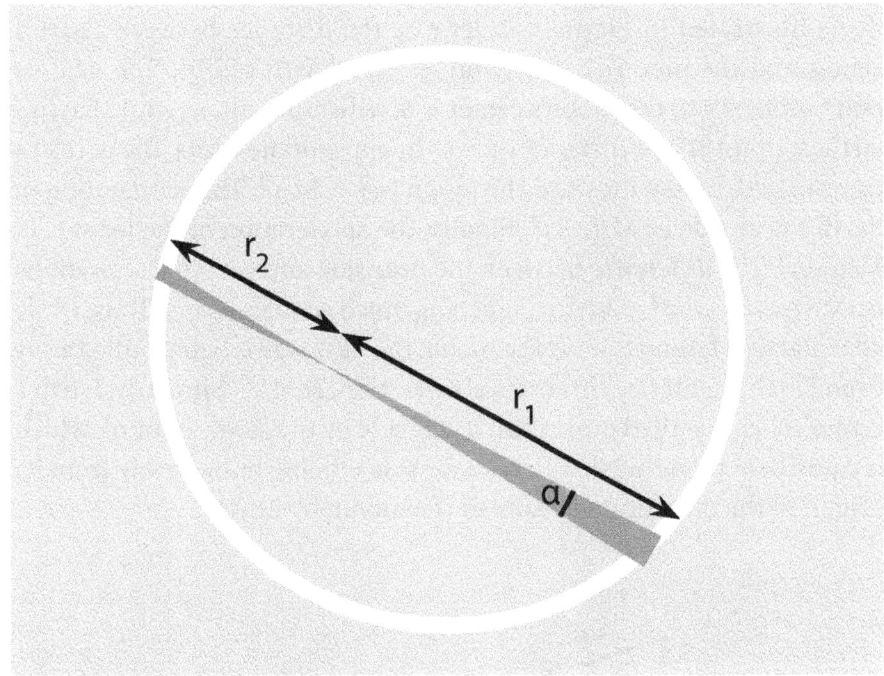


Figure C.1. This image shows a cross-section of a three dimensional shell. To compute the acceleration of gravity inside a thin, symmetric, massive shell, pick any point and place the tips of two cones at that point, with the cones aligned in opposite directions. The base of each cone covers a circular area of the shell. For each cone, the size of the shell area covered is proportional to the square of its r, its distance to the cone tip. But since the acceleration of gravity is inversely proportional to $r^2$, the factors of r cancel. Hence, the accelerations from each cone area are equal in magnitude but opposite in direction and sum to zero. The acceleration of gravity is zero throughout the interior of any symmetric shell.

Tides are a second effect of gravity, and are inversely proportional to the distance cubed: $1/r^3$. Earth's tides are due primarily to the moon and secondarily to the Sun. Initially, let's consider only the moon. The acceleration of gravity from the moon on different parts of Earth points in different directions and with different strengths, as we shall see. Here, we can still assume that the moon's mass is concentrated at a single point, but we have to reckon with Earth's actual size.

As illustrated in Figure C.2, let r be the distance between Earth's center and the moon's center, and let e be Earth radius. The nearest point on Earth to the moon's center is at a distance of r-e, while Earth's farthest point is at a distance of r+e. In appropriate units, the acceleration at Earth's center toward the moon is: $a = M/r^2$. The acceleration of Earth's near side is: $M/(r-e)^2$. Finally, the acceleration of the far side is: $M/(r+e)^2$. The ***difference*** between the near side and center accelerations is: $M/(r-e)^2 - M/r^2 = 2eM/r^3$, ignoring much smaller terms. Thus, while all of Earth is falling toward the moon, the near side is being pulled away from Earth's center by this extra acceleration, $2eM/r^3$. Similarly, Earth's center is being pulled more than its far side by the same amount, which is equivalent to saying that Earth's far side is being pulled away from its center, in the direction ***opposite*** to the moon, by $2eM/r^3$.

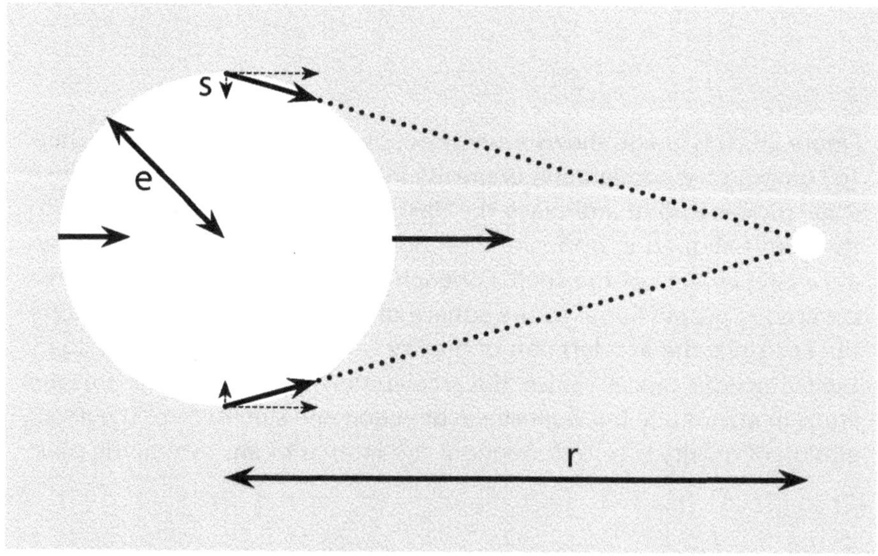


Figure C.2. Tidal forces due to the moon (small circle at right) are shown at various places on Earth (large circle at left). The Earth to moon distance is r and Earth's radius is e. Moon's pull is strongest at the nearest point on Earth and weakest at its farthest point. Midway on Earth's surface, the moon's pull is in a different direction, which we can separate into horizontal and vertical components, as shown above. The vertical component of pull, s, always points toward Earth's center.

Another tidal effect occurs at Earth's sides. Consider the circle around Earth that is exactly half way between the points nearest the moon and farthest from it. Everywhere on that circle the acceleration due to the moon's gravity points directly toward the moon's center, and everywhere that direction is slightly different from the direction of acceleration at Earth's center. These directions differ by an angle whose sine is e/r, and this difference always points toward Earth's center. We can see this by imagining that every point on this circle was free to fall independently toward the moon. Every point would collapse into the moon's center. In doing so, they would all converge to a single point, even though they started thousands of miles apart. Hence, all these acceleration differences point toward the center of this circle—Earth's center. Thus Earth's midsection, the zone halfway between the points nearest and farthest from the moon, is being squeezed inward with an acceleration that is e/r times $M/r^2$, the acceleration at Earth's center, which is $eM/r^3$.

All this results in Earth's tides. Water moves more easily than rock and thus responds more dramatically to these differing accelerations. Water is pulled toward the moon on Earth's near side, pulled away from the moon on Earth's far side, and pulled toward Earth's center in between near side and far side. Thus Earth always has high tides at two locations and one circumferential low tide, and as Earth rotates once daily, each spot on the shoreline experiences two high tides and two low tides every 24 hours and 50 minutes. If the moon didn't move, that time period would be 24 hours exactly. But since it orbits Earth once every 27.3 days, an extra 50 minutes of Earth's rotation is required to return the moon to the same position in our sky.

Note that all these tidal effects are proportional to $M/r^3$. Even though the Sun is 27 million times more massive than the moon, it is also 389 times farther from Earth, on average, and $389^3 = 59$ million. Thus the Sun's tidal effects on Earth are only 46% of the moon's, on average. When the Sun and moon align, they act in unison to produce our greatest tides, ***spring tides.*** When Sun and moon are separated in our sky by 90 degrees, their tidal effects conflict, producing minimal tides, ***neap tides***. Since the orbits of Earth and moon are not perfectly circular, and all heavenly bodies move, the magnitude of Earth's tides constantly change.

The third effect of gravity is time dilation. As described in chapter 10, in the section titled "Physics: Why Gravity Slows Time", a clock runs slower if it is closer to a gravitational source. In the example of a light pulse descending from the Leaning Tower of Pisa, photons gain energy (their frequency increases) as they lose gravitational potential energy. The frequency increase is proportional to the decrease in gravitational potential. Since gravity's acceleration is proportional to $1/r^2$, its potential energy is proportional to $1/r$. Mathematically, force times distance equals work energy, thus force is the derivative of energy with respect to distance, and $1/r^2$ is the derivative of $-1/r$. The magnitude of gravitational time dilation is $2M/r$, where M is the gravitating mass and r is the distance to its center, in appropriate units.

Gravity manifests itself in these three very different effects—acceleration, tidal forces, and time dilation—each varying with a different power of distance—the 3 r's of gravity.

Another book by Robert Piccioni

# *Everyone's Guide to Atoms, Einstein, and the Universe*

ISBN 978-0-9822780-7-9
Rated 4.94 Stars by Amazon Readers.
Find out why this is the highest-rated book in its category.
Winner of "*Best in Science*" in both USA Book News
and the 2010 International Book Awards.

The Midwest Book Review called this "The layman's guide to 20th and 21st century science...with straightforward comprehensible explanations... enthusiastically recommended to readers of all backgrounds as well as public libraries."

Robert says: "You don't have to be a great musician to appreciate great music. Similarly, you don't have to be a physics and math expert to appreciate the exciting discoveries and intriguing mysteries of our universe. If you can change money, you can understand Einstein."

With 150 illustrations, including 33 full-color images, *Everyone's Guide to Atoms, Einstein, and the Universe* explores modern physics, astronomy, and cosmology, covering elementary particles, antimatter, Special and General Relativity, Quantum Mechanics, black holes, supernovae, the Big Bang theory, dark matter, dark energy, the size and age of our universe, and where we humans fit in.

Another book by Robert Piccioni

# *Can Life Be Merely An Accident?*

ISBN 978-0-9822780-2-4
Rated 5 Stars by Amazon Readers.
Winner of "Best in Science"— Pinnacle Book Achievement Awards

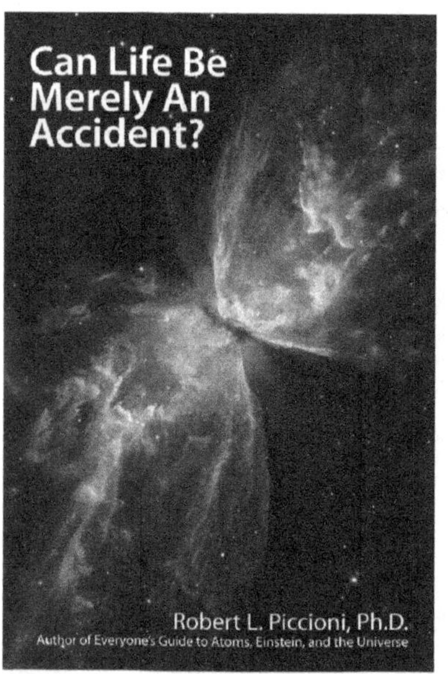


For decades, most scientists have believed that life began when the right molecules randomly collided in just the right way—that "time itself performed the miracles." Other scientists have rejected such an accidental origin as being as unlikely as "the assemblage of a 747 by a tornado whirling through a junkyard."

Dr. Piccioni's remarkable book is a scientific analysis of the many exacting requirements for the existence of any form of life—a universe with an exact geometry, the right atoms, a viable habitat, and an effective genetic code—and how extraordinarily improbable it is that these were fulfilled by random chance.

Robert demonstrates that our universe is tuned, with amazing precision, to support life, and that DNA is the most complex and precise structure in the entire cosmos. An explicit and comprehensive analysis concludes that life arising by random chance, through any currently known chemical or physical process, is extraordinarily unlikely—less likely that drawing the ace of spades 100,000 times in a row.